BEHAVIOR OF MARINE ANIMALS

Current Perspectives in Research

Volume 2: Vertebrates

BEHAVIOR OF MARINE ANIMALS

Current Perspectives in Research

Volume 1: Invertebrates
Volume 2: Vertebrates

BEHAVIOR OF MARINE ANIMALS

Current Perspectives in Research

Volume 2: Vertebrates

Edited by

Howard E. Winn

Professor of Oceanography and Zoology
Narragansett Marine Laboratory
University of Rhode Island
Kingston, Rhode Island

and

Bori L. Olla

Head, Behavior Department
U.S. Department of Commerce
National Oceanic and Atmospheric Administration
National Marine Fisheries Service
North Atlantic Coastal Fisheries Research Center
Laboratory for Environmental Relations of Fishes
Highlands, New Jersey

PLENUM PRESS • NEW YORK-LONDON • 1972

Library of Congress Catalog Card Number 79-167675

ISBN 0-306-37572-9

A Division of Plenum Publishing Corporation
227 West 17th Street, New York, N.Y. 10011

United Kingdom edition published by Plenum Press, London
A Division of Plenum Publishing Company, Ltd.
Davis House (4th Floor), 8 Scrubs Lane, Harlesden, London,
NW10 6SE, England

Printed in the United States of America

CONTRIBUTORS TO VOLUME 2

James F. Fish Naval Undersea Research and Development Center, San Diego, California

Richard N. Mariscal Department of Biological Science, Florida State University, Tallahassee, Florida

Arthur A. Myrberg, Jr. School of Marine and Atmospheric Sciences, University of Miami, Miami, Florida

Eugene L. Nakamura Eastern Gulf Sport Fisheries Marine Laboratory, National Marine Fisheries Service, Panama City, Florida (formerly of the Bureau of Commercial Fisheries Biological Laboratory, Honolulu, Hawaii)

Bori L. Olla U.S. Department of Commerce, National Oceanic and Atmospheric Administration, National Marine Fisheries Service, North Atlantic Coastal Fisheries Research Center, Laboratory for Environmental Relations of Fishes, Highland, New Jersey

Ronald J. Schusterman Stanford Research Institute, Menlo Park, California, and California State College at Hayward, California

Robert A. Stevenson, Jr. Rosenstiel School of Marine and Atmospheric Science, University of Miami, Miami, Florida

Anne L. Studholme U.S. Department of Commerce, National Oceanic and Atmospheric Administration, National Marine Fisheries Service, North Atlantic Coastal Fisheries Research Center, Laboratory for Environmental Relations of Fishes, Highland, New Jersey

Howard E. Winn Graduate School of Oceanography, University of Rhode Island, Kingston, Rhode Island

PREFACE

What have been brought together in these volumes are works representing a variety of modern quantitative studies on a select group of marine organisms. Some of the species studied here represent basic biological experimental subjects–in some cases, marine versions of the white rat and pigeon–that are being used for a wide range of studies. Other species studied were virtually unknown as experimental animals.

The authors have studied their animals in considerable depth, often in both the field and the laboratory. It is this cross reference between real life and the artificial but controlled conditions of the laboratory which gives us the necessary understanding, and ultimately the means, for improving our rapidly deteriorating environment, a must for man's survival, maintenance, and improvement of the quality of living standards.

A direct outgrowth of a AAAS symposium entitled "Recent Advances in the Behavior of Marine Organisms" held in December 1966, these volumes include a reasonable balance between review and original unpublished research.

Of the many persons who have made these volumes possible, we wish to especially thank Nancy Fish, Lois Winn, Mabel Trafford, and Deborah Brennan. The latter two accomplished most of the final editorial work. The personnel of Plenum Press were cooperative in all aspects of our relationship. Only the two editors are responsible for defects in the volumes. We believe the papers presented are significant and will be of importance to members of the scientific community.

Howard E. Winn
Bori L. Olla

INTRODUCTION

The study of animal behavior may be viewed in a simplified way as an examination of the potentialities of the nervous sytem with its corresponding effector system. The fundamental starting point, whether the environment be aquatic or terrestrial, is observing what an animal does. To the behaviorist, this means observing the external expressions of the nervous system and effector organs as manifested in how an animal reacts to a particular stimulus. It is important to bear in mind that the data resulting from the direct observation of behavioral events, especially those occurring in the aquatic milieu, often contain a high degree of inherent random variation. Therefore, it is essential that questions be formulated and defined carefully and the data be collected and managed in a way that permits valid conclusions.

Marine animals and questions about their behavior are the main concern of the small but representative number of works appearing in these volumes. Until recent years, the main deductions on behavior of marine animals have been the result of surface sightings and net and hook collections correlated with measurements of environmental parameters. The number of collections and measurements depended, as it does today, on the ingenuity and budget of the individual scientist as well as the state of instrument development. Observations on marine animals in their natural environment have been encumbered to some degree by the fact that the observer has been physically unable to venture beneath the sea for more than a brief time. Over the past 15–20 years, the refinement of the self-contained breathing apparatus has extended the time and versatility of underwater observation. Improvements in manned submersibles, underwater television, and an array of remote sensing devices, have also contributed greatly to the opportunities for *in situ* study. Because it is not possible to control the myriad environmental influences in the sea, it is important to measure simultaneously various physical and biological parameters both above and below the sea. Subsequent correlation of these parameters with behavioral events permits meaningful interpretations and conclusions on cause and effect relations.

Relating laboratory findings to what is known of the animal's behavior in the sea is often a necessary step in interpreting and evaluating results. With all of its problems, the laboratory, because of the potentiality for precise control of most environmental cues, permits judgments about behavioral influences with less equivocation than is possible in the field. The ideal situation is one in which observations take place in both the field and laboratory with a continuous cross formulation of questions and integration of results.

There are certain requirements basic to laboratory study. An aquatic medium must be provided which closely approximates or duplicates natural sea water. Obvious parameters such as temperature, salinity, and various chemical balances must be constantly monitored and controlled to avoid variations in behavior caused by changes in the water medium. Confinement, as is the case with almost any vertebrate, be it terrestrial or aquatic, will often modify behavior in an unknown manner. Thus it is necessary to compare field and laboratory findings to determine what are real and what are laboratory-induced events.

The aim of this volume is to present a variety of approaches and methods of analysis of both field and laboratory studies at various phylogenetic levels. Our goal is to show some small segment of the work that is currently being done in the area of marine animal behavior. The reader should, upon perusing the various chapters, note the interrelationships between organisms even though they are at different phylogenetic levels, and note the behavioral adaptations that permit the various species to be in harmony with the very specialized environment of the sea.

Most of the studies contained in this volume were performed on animals that live on the fringes of the sea. It is true that the accessibility of these species to the researcher makes them prime candidates for study. But more important than their availability is the fact that these organisms living in the estuaries and inshore areas are residing in the richest marine habitat. In contrast, the open sea is a veritable desert. Unfortunately, these inshore areas and their resident flora and fauna are highly vulnerable to man-made changes and destruction in the name of progress. Defining the destruction in a legal sense has been difficult because often no studies are undertaken until after the greatest damage has been done. Studies which concentrate on establishing normal patterns of behavior are necessary to predict the outcome of particular man-caused modifications. Herein lies a valuable tool and asset in our attempt to salvage the environment, a tool that has not been adequately used.

Howard E. Winn
Bori L. Olla

CONTENTS OF VOLUME 2

CONTENTS OF VOLUME 1

Chapter 3

Ritualization in Marine Crustacea

by Brian A. Hazlett

Chapter 4

Aggressive Behavior in Stomatopods and the Use of Information Theory in the Analysis of Animal Communication

by Hugh Dingle

Volume 2: Vertebrates

Chapter 6

DEVELOPMENT AND USES OF FACILITIES FOR STUDYING TUNA BEHAVIOR

Eugene L. Nakamura*

Bureau of Commercial Fisheries Biological Laboratory
Honolulu, Hawaii

I. INTRODUCTION

Tunas are pelagic fishes of several genera† belonging to the family Scombridae. They are fished commercially and recreationally in the temperate and tropical seas of the world. Sizes of adult tunas range from less than a kilogram (e.g., *Auxis rochei*) to over 400 kg (e.g., *Thunnus thynnus*).

Accumulation of knowledge on the behavior of tunas was slow and haphazard before the 1950s, but has accelerated since then. In 1951, the HL (Bureau of Commercial Fisheries Biological Laboratory in Honolulu, Hawaii), known then as the Pacific Oceanic Fishery Investigations, contracted with researchers at the University of Hawaii to undertake studies of the behavioral responses of tunas to assorted stimuli. These experiments required the establishment of tunas in captivity. This feat was successfully accomplished for the first time in 1951. In the following years, techniques and methods for capturing and maintaining tunas in captivity were improved, and tanks were constructed especially for studying tuna behavior. In the mid-1950s, development of facilities to observe tunas underwater at sea was started. Significant advances in studies of tuna behavior have been made owing to observations and experiments made at sea and ashore with these facilities.

Reviews of tuna behavior have been made by several authors. Tester

* Present address: Eastern Gulf Sport Fisheries Marine Laboratory, National Marine Fisheries Service,PanamaCity,Florida.

† Species of the following genera have been included as tunas in this paper: *Thunnus*, *Euthynnus*, *Katsuwonus*, *Auxis*, and *Sarda*.

(1959) summarized the studies he and his colleagues conducted from 1951 through 1956 on the responses of tunas to stimuli. Magnuson (1963) reviewed the literature on tuna behavior and physiology. Behavioral studies that were being conducted at HL were described by Magnuson (1964*b*). A popularized account of the work at HL was given by Manar (1965). Nakamura (1969*a*) has summarized observations on tuna behavior made at sea.

The purpose of this chapter is to present a history of the development of the techniques and facilities for studying tuna behavior, a statement of the problems and questions that prompted their development, and an account of the uses to which they were put. Most of the facilities and methods were developed at HL.

II. OBSERVATIONS AND RESEARCH AT SEA

Observations of tuna behavior at sea before 1957 were made only when opportunity afforded, usually from the deck of a boat—particularly boats used in fishing for tunas by pole and line. In this method, tunas are attracted to the boat by throwing overboard live bait, which is kept aboard in bait-wells. The tunas are then caught with pole and line (Yoshida, 1966). Various aspects of feeding, schooling, swimming, reactions to stimuli, behavior in relation to environmental features, and associations with other organisms and objects have been learned by observing from the decks of boats (Nakamura, 1969*a*).

The need to learn more about tuna behavior by observing tunas in their own environment became more and more apparent to biologists at HL in the early 1950s. Answers to such questions as the effectiveness of various species of bait, the possibility of controlling tuna behavior by manipulating fishing conditions, reactions elicited in tunas by actions of the bait, the efficacy of gear, the necessity for live bait, were difficult to obtain without going below the sea surface to observe.

The first attempt to observe tunas underwater during fishing was made in 1957. A steel ladder was lowered and secured to the side of HL's research ship *Charles H. Gilbert*. An observer descended, so that his head was about 0.5 m below the surface. He wore a regulator with an air hose attached to a cylinder of compressed air on the deck of the ship. He faced aft and observed tunas which had been attracted to the stern of the ship with live bait. This method had some notable disadvantages: the ship's speed had to be greatly reduced, observations could be made only in calm water, the observer had a difficult time keeping his purchase on the ladder and on his face mask and mouthpiece against the current, and the observer's view was often obstructed by bubbles (Strasburg and Yuen, 1960*b*). Subsequently, a protective shield was built aroung the ladder, providing sufficient freedom to photograph.

Despite the shield, the ladder was unsatisfactory. The mouthpiece and regulator were bothersome, and communication with the ship's personnel was crude, to say the least (prearrangements were made to signal changes of events by pounding on the ship's hull). Use of this ladder did convince us, however, that observing tuna behavior underwater was feasible. Plans were therefore developed for a structure that would permit the observer to remain in air but still be below the sea surface and that would permit voice communication.

A retractable overside caisson (Fig. 1) was built in 1958 (Strasburg

Fig. 1. The retractable caisson.

and Yuen, 1960*b*). The caisson was a tube about 3.5 m long. It was large enough to permit one person to observe and photograph through ports about 1–1.5 m below the surface of the sea. When not in use, the caisson was raised out of the water with block and tackle. Later, the tube was extended to prevent water from splashing in over the top, the ports were enlarged to permit a wider field of visibility, and a winch was installed for lowering and raising it.

Use of the caisson was described by Strasburg and Yuen (1960*a*) and Strasburg (1959). They observed the behavior of skipjack tuna (*Katsuwonus pelamis*) under varying fishing conditions, such as when water sprays were turned on and off, when live and dead bait were chummed, when the rate of chumming was altered, when different species of bait were chummed, when body fluids from dead skipjack tuna were poured into the water, and when noise was presented to the tuna. Changes in schooling behavior, feeding behavior, coloration, and abundance were recorded during the experiments.

A serious disadvantage in using the caisson was the poor visibility in

Fig. 2. The stern observation chamber of the *Charles H. Gilbert.*

Fig. 3. School of skipjack tuna (*Katsuwonus pelamis*) photographed from stern chamber.

the after direction caused by bubbles in its wake. To overcome this problem, installation of underwater viewing ports in the hull of the *Gilbert* was proposed. The location of these ports as far astern as practicable seemed desirable, since fishing was at the stern.

After problems involving reductions of fuel storage, effect on ship's speed and steerage, turbulence, etc., were considered, an observation chamber was installed in 1959 on the port quarter just forward of the propeller (Fig. 2). Dimensions and details of construction have been provided by Akana *et al.* (1960), Strasburg and Yuen (1960*b*), and Mann (1961).

The stern chamber consists of a streamlined semicylindrical shell. Two ports permit observations from about 2 m below the sea surface directly below one of the fishing racks. Although the chamber can accommodate two persons, only one can observe comfortably in the semicylinder. The entrance is on the after bridge deck. A blower system provides ventilation. Communication is via a sound-powered telephone with headsets. The headsets help to block out much of the noise generated by the ship's machinery. Visibility is excellent and unimpeded by bubbles (Fig. 3).

The stern chamber was used during additional experiments on variation of fishing conditions. Strasburg (1961) noted a relation between diving frequency of schools of skipjack tuna and the presence of certain species of prey in stomachs of fish caught from these schools. Yuen (1969) observed catch rates, rates of attack on bait, and numbers of tunas attracted to the vessel during the experimental fishing. The stern chamber has also been used to study the behavior of different species of bait and the behavior of tunas in response to the behavior of the bait (unpublished).

Schooling behavior, swimming behavior, and coloration of tunas have been observed and recorded from the stern chamber. Banded color phases of skipjack tuna (and also dolphin, *Coryphaena hippurus*) were observed and photographed by Strasburg and Marr (1961). Motion pictures taken from the chamber of yellowfin (*Thunnus albacares*) and skipjack tunas were analyzed to study schooling behavior (Yuen, 1963) and to determine swimming speeds (Yuen, 1966).

Although the stern observation chamber permits excellent observations of tuna behavior during fishing, it does not permit observations of tuna schools ahead of the ship. An observation chamber in the bow of the ship seemed desirable for observing while sailing up to and through schools of tuna or while following them.

A bow chamber (Fig. 4) was installed in the *Gilbert* in 1960 (Akana *et al.*, 1960; Strasburg and Yuen, 1960*b*; Mann, 1961). To do so, the bottom of the bow was reshaped into a bulbous form to provide space for an observer.

Fig. 4. The bow observation chamber of the *Charles H. Gilbert*.

Three ports were provided for viewing straight ahead and to both sides from about 2 m below the waterline.

The bow chamber comfortably accommodates one observer. Access is from the forecastle. The chamber is ventilated by a forced-draft system. As in the stern chamber, a sound-powered telephone is used for communication.

Soon after the bow chamber was constructed, Yuen (1961) had opportunities to observe and film bow wave riding by porpoise (*Tursiops* sp.). The method of bow wave riding by porpoises was the subject of a controversy among scientists at the time. Yuen settled the matter by describing "the method the (porpoises) themselves seem to consider proper."

The bow chamber has been used to observe the effectiveness of gill nets after skipjack tuna were attracted toward the net with live bait (unpublished). Most of the tuna evaded the net, which was clearly visible from about 20 m. A few skipjack tuna were seen to swim into the net, become gilled, struggle a few seconds, and then succumb.

The ports in both the stern and bow chambers require periodic cleaning, but, in general, visibility in near-surface waters during the day is limited only by water clarity and the endurance and will of the observer. When the ship is heading into moderate to rough seas, the bow chamber often rises above the surface, causing discomfort to all but the most hardy.

Use of the stern and bow chambers in the *Gilbert* does have limitations. Tunas cannot be observed at night, and they cannot be observed at depths greater than about 25 m. Tunas cannot be followed, because their maneuverability is greater than that of the ship. Several methods were considered to overcome these problems.

Underwater television was tested in 1959 (Strasburg and Yuen, 1960*b*). A closed-circuit set was installed on the *Gilbert*. A television camera was set in the stern chamber, and another was suspended in the water. Observations of tuna behavior with television were found to be inferior to observations with the human eye under good light conditions. Under poor light, television provided good observations owing to better contrast. The major difficulty with this method was our inability to shift quickly the field of view. The use of television therefore was abandoned.

In 1959, responses to inquiries sent to manufacturers of submersible vehicles indicated that a submarine for studying various biological and oceanographic problems in addition to tuna behavior appeared practicable. Much thought and effort were put forth in the early 1960s on the uses, on the scientific and operational requirements, and on the design of a research submarine. In 1964, an industrial company was contracted by HL to undertake a feasibility and conceptual-design study. The vessel was required to travel as fast as 20 knots while submerged, remain submerged for as long as 6 weeks, dive as deep as 300 m, and have an operational range of 40,000 km. The specified requirements and design of the research submarine (Fig. 5),

Fig. 5. Plastic model of a proposed nuclear research submarine.

which called for nuclear power, were found to be feasible. One of the great advantages of a submarine is its independence of weather and sea state. Observations are possible in relative comfort while the submarine is submerged despite inclement weather or rough seas. Uses, design, requirements, and facilities of the submarine were presented by Strasburg (1965).

In 1965, HL chartered the *Asherah* (Fig. 6), a submersible about 5 m long, to gain experience in operating and conducting missions with an under water craft. The *Asherah* carried one observer in addition to the pilot. It was operated in the lee of the island of Oahu, Hawaii. Skipjack tuna and kawakawa (*Euthynnus affinis*) were seen on six of the 50 dives with the *Asherah* at depths as great as 152 m (Strasburg, 1966; Strasburg *et al.*, 1968). The tunas were observed preying upon forage fish, but because of the limited speed and maneuverability of the *Asherah* the tunas could not be followed.

A charter of a larger submarine to gain further experience was planned, but owing to insufficient funds the plan was cancelled. Early in 1967, plans for constructing the nuclear research submarine were shelved for the same reason.

Acoustic means of studying tuna behavior at sea were also considered. Acoustic devices could be used to locate and track tunas at night as well as during the day and at depths and distances beyond visual range. Echo-sounders had been used to study movements and speeds of large, deep-swimming tunas (several species of *Thunnus*) by Nishimura (1963, 1966), but for faster-swimming tunas in waters near the surface conventional sonars which transmit sound impulses at fixed intervals and at fixed frequencies were inadequate for tracking. Such fast-swimming fishes could easily escape detection between transmissions. A sonar that could keep tunas under continuous surveillance was desired.

A CTFM (continuous-transmission, frequency-modulated) sonar (Fig. 7) was installed in HL's research ship *Townsend Cromwell* in 1966. Additional equipment has been installed so that pertinent data can be recorded

a

b

Fig. 6. The submarine *Asherah*. (**a**) Loading for transport. (**b**) Boarding at sea.

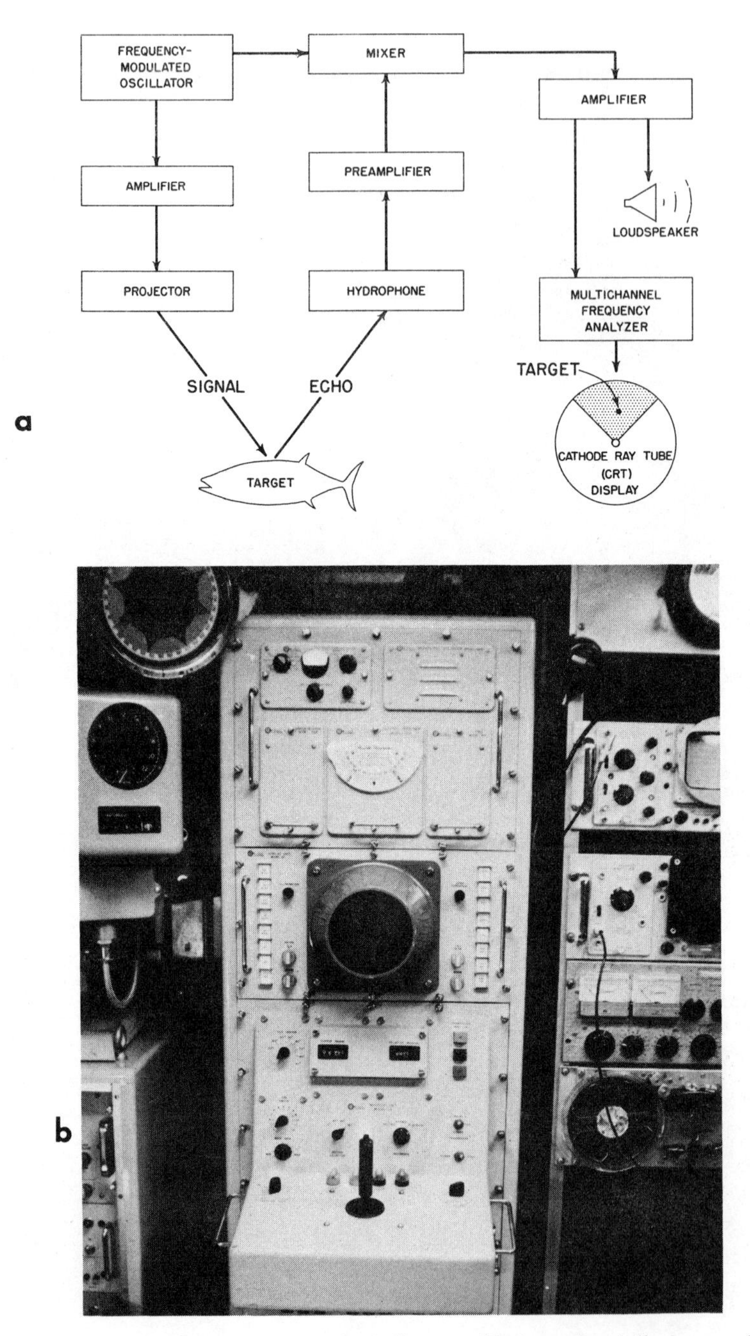

Fig. 7. The CTFM sonar. **(a)** Block diagram. **(b)** Control and display panels.

on a 14-channel magnetic tape recorder. The sonar has two operational modes: a search mode with frequencies modulated from 52 to 32 kHz for targets up to 1600 m, and a classify mode with frequencies modulated from 290 to 260 kHz for high resolution at short ranges (100 m). The data from the CTFM sonar are recorded in analog form on magnetic tapes. The tapes are brought ashore, where the data are digitized and then analyzed with the aid of computers. Descriptions of the sonar units and of the operations have been given by Yuen (1967, 1968).

The CTFM sonar, which projects and receives acoustic energy continuously, thereby lessening the possibility of losing a fast-moving target, has been used to observe movements of tuna schools. Diving and swimming behavior of yellowfin and skipjack tunas in relation to their size and to temperatures and depths have been investigated (Yuen, 1968). Schools of skipjack tuna have been followed for many hours continuously while observing their behavior in relation to time and space. These and other studies, including use of ultrasonic tags with the sonar, are in progress.

Hester (1969) attempted to use a CTFM sonar to identify tunas and other species and to determine their sizes by analyzing shifts in frequencies (Doppler shifts) caused by motions of the fishes. The loss of echo strength when fish turned toward or away from the sonar made this method impractical.

Another method used in 1967 for underwater observations was a sea sled towed by the *Gilbert* (Anonymous, 1967). The sea sled was manned by

Fig. 8. Photographing from sea sled.

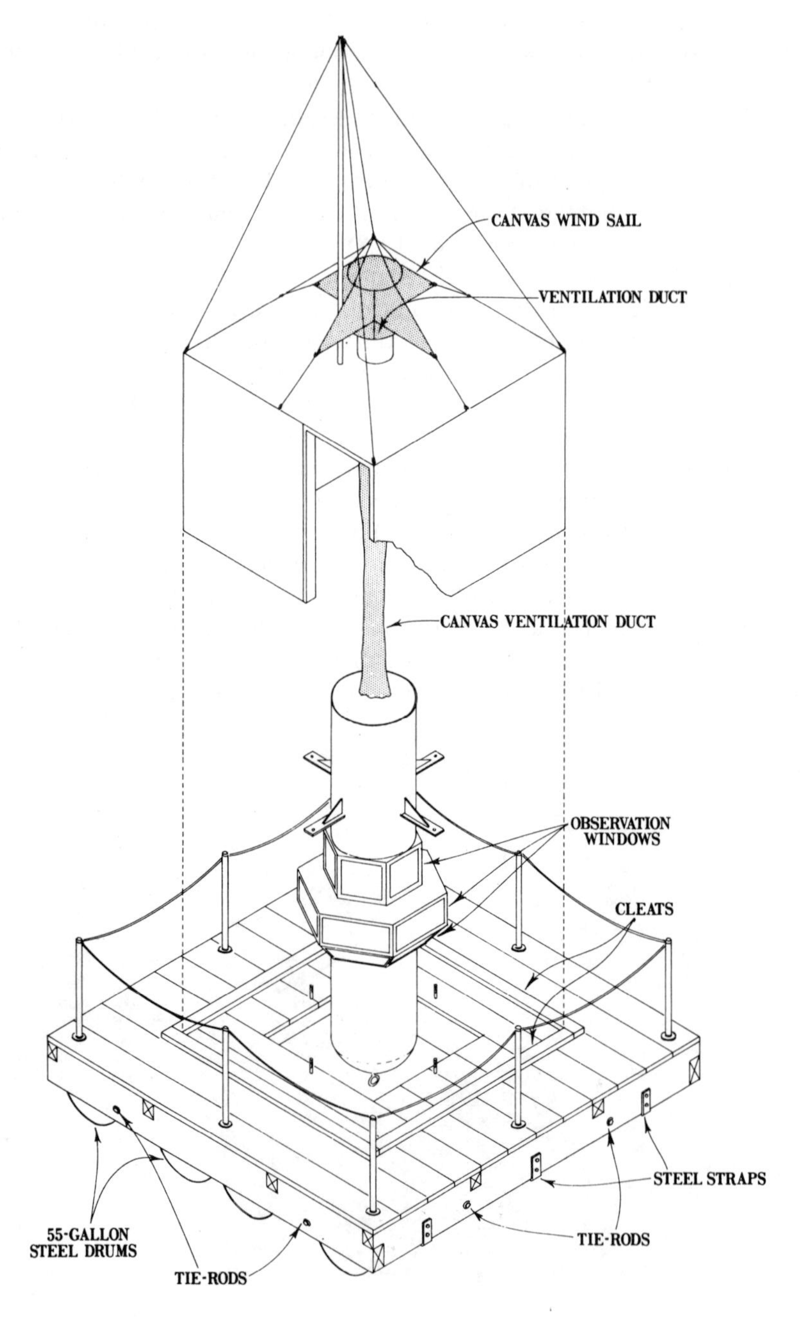

Fig. 9. The raft *Nenue*. **(a)** Diagram.

two divers equipped with aqualungs (Fig. 8). Communication with the ship was conducted by prearranged signals through a buzzer system. The sled was constructed at the Bureau of Commercial Fisheries Exploratory Fishing and Gear Research Base, Seattle, Washington, by fitting diving planes, windshield, and other equipment to a basket litter.

The behavior of baitfish and the behavior of skipjack tuna preying upon the bait were observed and photographed from the sled, which was towed at about 2 knots at distances up to 50 m behind the ship and as deep as 15 m. The sled proved to be impractical, because by the time it was launched and in operation, the tuna school usually had departed. However, the first underwater observations of courting behavior of skipjack tuna were made (Iversen *et al.*, 1970) before use of the sea sled was discontinued.

During the late 1950s and early 1960s, a number of papers appeared describing parasite-picking and cleaning behavior by fishes and crustaceans in reef communities. Members of HL began to wonder if this phenomenon existed in pelagic environments also. Tunas were often found with ecto-parasitic copepods. Which organisms, if any, picked parasites off large

Fig. 9. **(b)** Observer in cylinder.

pelagic fishes? Tunas were often found in association with drifting objects at sea. These drifting objects almost always had small fishes under them. Could these small fishes act as cleaners for tunas and other pelagic predators?

In an attempt to answer these questions, a raft with observational facilities (Fig. 9) was built in 1962 (Gooding, 1965). The square raft had about 4 m to a side with an aperture in the center through which a steel cylinder was suspended. The cylinder was provided with windows permitting underwater observations in all directions except immediately below. The raft was named *Nenue*, which is the Hawaiian name of *Kyphosus cinerascens*, a fish commonly found under floating objects. The *Nenue* was manned by two persons, who took turns observing in the suspended cylinder. The raft was occupied only during daylight hours. Each evening the observers left the *Nenue* to return to the tending research ship.

The *Nenue* proved to be highly seaworthy, more so than some of the observers. (One observer was removed after 15 min aboard the raft; he had threatened to jump off and swim ashore if he were not taken off.) Many interesting observations and photographs (Fig. 10) were obtained of animals that visited and accumulated under and around the raft. After 13 drifts, 11 in the lee of the island of Hawaii and two south of Hawaii near the Equator, use of the *Nenue* was terminated. Although skipjack and yellowfin tunas were seen, no noteworthy behavior of these tunas was observed (Gooding, 1964; Gooding and Magnuson, 1967). The question concerning cleaning of tunas remains unanswered.

Observations of tunas under floating objects have also been made by Hunter and Mitchell (1967, 1968). They made their observations by diving near floating objects which they found or which they built and placed in the water. Courting behavior and duration of association of black skipjack tuna (*Euthynnus lineatus*) with the floating objects were noted.

III. OBSERVATIONS AND RESEARCH ASHORE

The previously mentioned contract between HL and the University of Hawaii called for determining the reactions of tunas to various chemical, visual, and physical stimuli in the hope that the results might lead to new methods of catching tunas or to improvements in existing methods. To conduct controlled experiments, tunas had to be established and maintained in captivity. Since tunas had never before been held captive for experimental studies, development of methods and techniques for doing so became the first objective.

In 1951, two species of tunas, kawakawa and yellowfin tuna ranging

Fig. 10. Fishes seen from observation cylinder of *Nenue*. (**a**) Whale shark (*Rhincodon typus*) with unidentified remora. (**b**) Whitetip shark (*Carcharhinus longimanus*) with pilotfish (*Naucrates ductor*).

c

d

Fig. 10 (*continued*). (**c**) Wahoo (*Acanthocybium solandri*). (**d**) Juvenile dolphin (*Corpyhaena* sp.).

Fig. 10 (*continued*). (**e**) Freckled driftfish (*Psenes cyanophrys*).
(**f**) Adult dolphin (*Coryphaena hippurus*).

from 37 to 58 cm long, were successfully established in a small concrete tank and a large pond after initial failure in a small pond (Tester, 1952). Tunas were caught by surface trolling. They were transported to shore in the baitwell of the boat. A dip net was used to transfer fish from the baitwell to the pond or tank.

The concrete tank was about 11 m long, 3 m wide, and 1.2 m deep. Salt water was supplied by pumping from a channel. Baffles had been used to round the corners, because the tunas tended to swim along the walls of the tank and had difficulty in turning, especially if they were swimming rapidly when they reached a right-angle corner. The baffles eliminated the sharp corners. (The corners in the baitwell of the boat also were rounded with baffles.)

The pond had concrete walls built along a channel dredged out of a coral reef. The length was about 115 m, the width about 21 m. The depth of the pond varied from less than 1 m to about 4 m. Salt water was circulated by tidal currents through screened gates.

Kawakawa survived in the pond for over 2 years. Yellowfin tuna were kept in captivity for shorter periods. Attempts to establish frigate mackerel (*Auxis thazard*) and skipjack tuna in captivity were unsuccessful. Dolphins were also successfully established in captivity. Only the kawakawa and yellowfin tuna were used in experiments (Tester, 1952, 1959).

Responses of kawakawa and yellowfin tuna to visual, auditory, chemical, and electrical stimuli were observed in the pond and tank. Hsiao (1952) studied responses to white and colored lights at various intensities, and to continuous and interrupted white light. Hsiao and Tester (1955) conducted experiments on responses to moving lures of various colors. Some of their experiments were conducted when food extracts were introduced into the water. Experiments on attraction and repelling of tunas by sounds of various frequencies were conducted by Miyake (1952). He also investigated sound production by tunas. Responses to chemical stimuli were studied by Van Weel (1952) and Tester *et al.* (1954, 1955) by use of food extracts, water in which bait had been living, and assorted organic and inorganic chemicals. Tester *et al.* (1954) made ersatz baits by impregnating gelatin and macaroni with extracts of fish and squid, and tested them on both captive and wild tunas. Responses to electrical stimuli were studied briefly by Miyake and Steiger (1957). All of these experiments were summarized by Tester (1959).

In the late 1950s, facilities for holding and observing tunas ashore were developed at HL to complement the facilities developed to observe tunas at sea. A salt water well was dug in 1958 near the waterfront to get clean sea water—salt water which had undergone natural filtration and thus was relatively free of pollutants and marine organisms. Because the water lacked oxygen, an aerator consisting of layers of perforated trays was built (Strasburg, 1964). A circular, plastic-lined swimming pool (Fig. 11) about 7 m in diameter and 1.2 m deep was set up to hold the captive tunas (Nakamura, 1960).

Initial attempts to establish skipjack tuna in the pool failed. The tuna were captured by pole and line after they had been attracted to the ship with live bait. The fish were pulled out of the water, held by the fishermen while being carefully unhooked, and then placed in the baitwell. In port, the tuna were dip-netted, carried to the pool in a net (wet towels and plastic bags containing seawater were also tried), and released. None survived for more than a few hours. To counteract the excited actions exhibited by skipjack tuna, a drug (thorazine) was injected intramuscularly in some immediately after capture to tranquilize them. The additional excitation from handling while the injection was administered probably offset any tranquilizing effect the drug may have had. Again, none survived for longer than a few hours after release in the pool. All of the skipjack tuna incurred contusions, abrasions, loss of mucus, as indicated by discolored areas of the body, and nervous excitation from handling during these procedures. To prevent these effects, a method which eliminated handling was developed in 1959.

An elliptical steel tank 2.4 m long, 1.8 m wide, and 0.6 m deep was built to transport the tunas from sea to shore. When a fish was hooked, the fishermen swung the pole over the tank, quickly lowered the fish into the tank, and slackened the line to permit the fish to shake the barbless

Fig. 11. Circular, plastic-lined swimming pool for holding tunas.

hook out of its mouth. Once inside the tank, which was continuously provided with seawater, the fishes swam along the walls, but since the tank had no corners they were not required to make sharp turns. When the ship returned to shore, the tank was removed with a crane and immersed in the pool, and the fishes were allowed to swim out through a side hatch. Skipjack tuna brought ashore by this method did not have discolored areas, swam less excitedly and more slowly, and were successfully established in captivity. General observations were made on feeding behavior, coloration, swimming and schooling behavior, and recovery from injuries on these captive skipjack tuna (Nakamura, 1962).

After methods and facilities for establishing skipjack tuna and other species in captivity had been developed, additional experimental and observational tanks and pools were built. To determine visual capabilities of tunas, a U-shaped tank 15.9 m long, 4.9 m wide, and 1.2 m deep was built in 1960. The tank (Fig. 12) was enclosed in a Quonset hut to control light. Also, the Quonset hut facilitated observations by preventing rippling of the water surface by winds. Clear glass windows were placed in the ends of the tank. The observer conducted experiments from observation booths placed over the ends of the arms.

Visual acuities of skipjack tuna, kawakawa, and yellowfin tuna were measured in the tank in the Quonset hut. The method involved training tunas to respond differently to horizontal and vertical stripes which were projected onto an opal glass plate at a window in the tank (Nakamura, 1964*a*, 1968 1969*b*).

In 1961, a larger well was dug, and a new modified aerator was constructed (Nakamura, 1964*b*).

Another circular swimming pool was set up in 1962 for conducting experiments on hearing abilities of tunas. The pool was lined with mats made of rubberized pig and horse bristles to provide acoustic insulation. Experiments were conducted from an elevated observation booth built alongside the pool (Fig. 13). Training methods were used again to determine frequencies of sound perceived by yellowfin tuna and kawakawa (Iversen, 1967 1969).

Since the circular swimming pool was found to be most useful for observing tuna behavior, an array of pools was set up in 1962. Six pools were arranged in a ring around an observation tower (Fig. 14). An aerator, which sprayed water through small holes, was built in the center of the observation tower. Water flowed from the aerator into the pools and then out through a drain in the center of each pool. Since seawater from the well was plentiful, recirculation was unnecessary. Windows in the tower permitted views of each tank. Three small huts were constructed, one between each

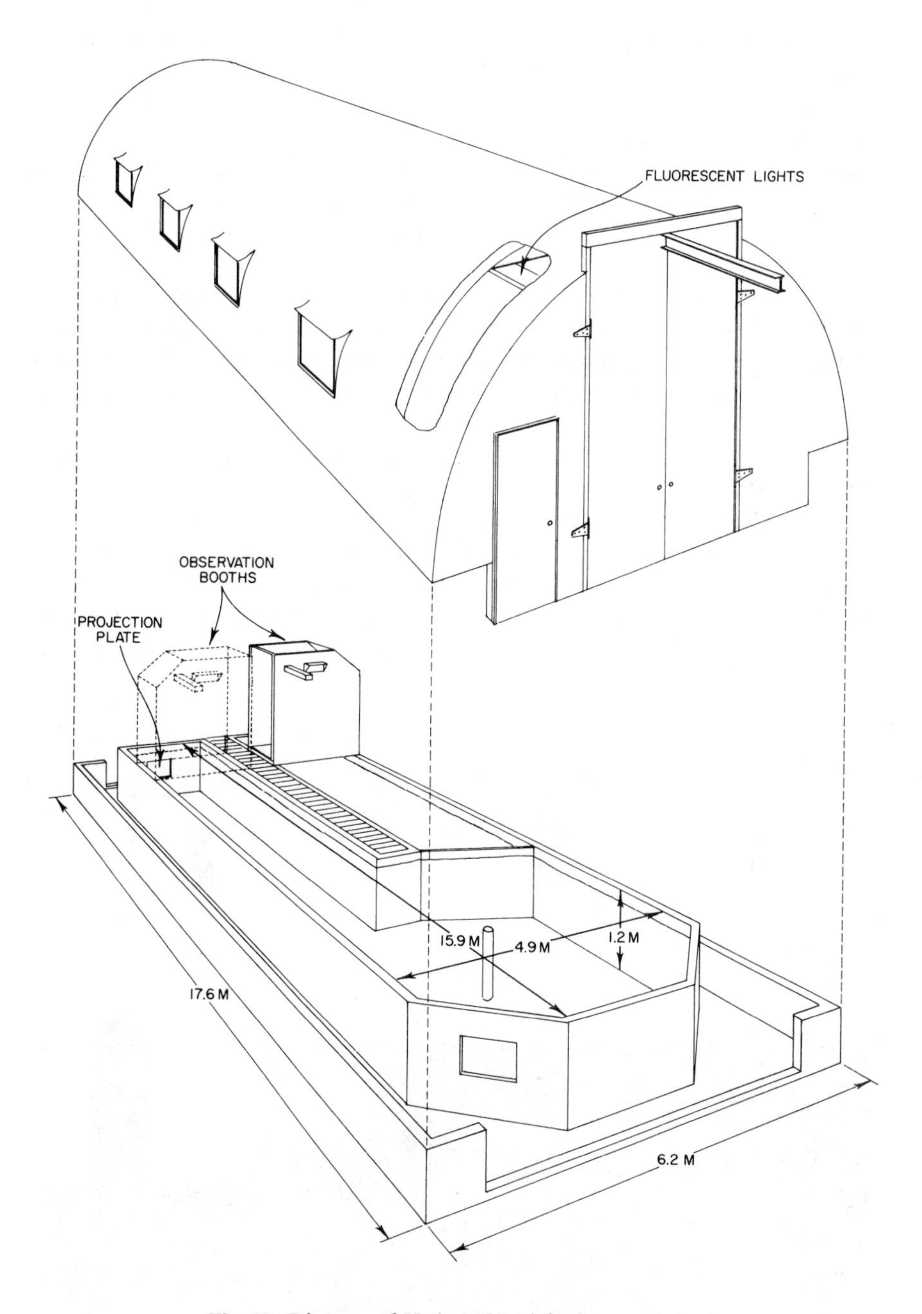

Fig. 12. Diagram of U-shaped tank in Quonset hut.

pair of pools. Windows in the huts permitted observations from below the water surface (Magnuson, 1965).

Awnings were placed over the six-pool complex and the acoustic pool in 1965 to reduce the amount of sunlight entering them. Because a diatom of the genus *Melosira* grew rapidly and profusely in pools open to the sun, the pools had to be cleaned twice a week. After the awnings were placed, the frequency of cleaning was reduced to once every 1 or 2 weeks.

The need to obtain more tunas per trip plus problems concerning the weight and maintenance of the steel transfer tank led to the construction in 1964 of five fiberglass transfer tanks. The new tanks (Fig. 15) were similar to the steel tank in size and design except for removable tops. The fish were released by removing the top and overturning the fiberglass tank in the pool (Nakamura, 1966).

Several species of tunas (Fig. 16) have been kept in our pools at HL. The length of survival has varied with species. Kawakawa have been kept for years, but frigate mackerels (*Auxis thazard* and *A. rochei*) have not survived for longer than a month. Other tunas, such as yellowfin tuna, bigeye tuna (*Thunnus obesus*), and skipjack tuna, have survived for months—generally until they were either sacrificed for experimental reasons, lost owing to accidents, or poached (one of the few people ever to catch tunas by

Fig. 13. Pool and observation tower for acoustic experiments.

Fig. 14. Array of six swimming pools around an observation tower with small observation huts between each pair of pools.

spearfishing was a youngster who sneaked into our pools in the evenings). Skipjack tuna, however, have not survived longer than 6 months. Other pelagic fishes, such as the dolphin and the rainbow runner (*Elagatis bipinnulatus*), have also been successfully established in our pools.

Small specimens are sought in fishing for tunas to establish in captivity. The size of the transfer tanks and the pools has favored smaller fish. Also, more small tunas can be placed in a transfer tank and in a swimming pool.

Behavior of tunas in HL's pools has been observed visually from the central observation tower, from the small hut adjacent to the pools (Magnuson, 1965), and from the pool's edge. Cine and still photography and other special equipment have been used in experiments conducted in these pools. High-speed cinematography was used by Walters (1966) to investigate the mechanics of feeding by kawakawa. Schooling behavior of kawakawa has been observed by Cahn (1967). Magnuson (1964*a*, 1966*a,b*, 1969*b*) has studied swimming behavior of kawakawa in relation to respiration and feeding and to hydrostatic equilibrium, and Magnuson (1970) has analyzed the hydrostatic functions of the appendages of kawakawa. He has also determined digestion rates in skipjack tuna (Magnuson, 1969*a*). Changes in color and in body markings have been noted in several species

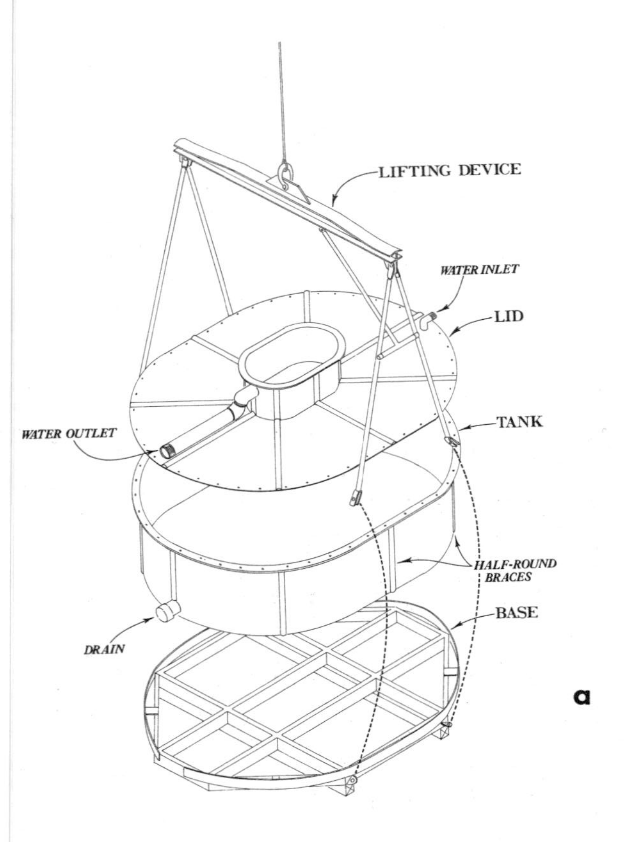

Fig. 15. Fiberglass transfer tank. (**a**) Diagram. (**b**) Placing skipjack tuna (*Katsuwonus pelamis*) into transfer tank.

Fig. 16. Tunas kept in captivity. (**a**) Skipjack tuna (*Katsuwonus pelamis*). (**b**) Bigeye tuna (*Thunnus obesus*).

Fig. 16 (*continued*). (**c**) Kawakawa (*Euthynnus affinis*). (**d**) Kawakawa preying upon a school of baitfish (*Stolephorus purpureus*).

of tunas (Nakamura, 1962; Nakamura and Magnuson, 1965) and also in the dolphin (Murchison and Magnuson, 1966). Some preliminary experiments (unpublished) on olfaction and oxygen tolerance in tunas have been conducted in the pools and have indicated sensitivities and awareness by tunas to dilute concentrations of aquatic odors and to changes in oxygen tension.

While the behavior of captive tunas was observed, questions of sensory capabilities arose which led to studies of the morphology and histology of several organs. The ability of tunas to perceive objects thrown to them before the objects hit the water, even when the surface was rippled, led to examination of the gross morphology and retinal histology of the eyes of several species of tunas (Matthews, unpublished work cited by Tester, 1959). The nature of the small openings of the anterior nares and slitlike posterior nares led to studies of the mechanics of water flow and the morphology and histology of the olfactory organ of skipjack tuna (Gooding, 1963). Questions concerning the silvery tongue of skipjack tuna led to histological examination for taste buds, which were found. A layer of guanine crystals was responsible for the silvery appearance of the tongue (unpublished).

The availability of live tunas at HL has permitted several investigators to carry out anatomical and physiological studies requiring live or fresh specimens. Physiology of muscles of skipjack tuna has been investigated by several workers (Chung *et al.*, 1967; Rayner and Keenan, 1967; Sather and Rogers, 1967). Anatomical and electrophysiological studies of the nerve in the trunk lateral line of skipjack tuna have been made by J. Suckling (1967) and E. Suckling (1967). Oxygen consumption in muscles of skipjack and bigeye tunas has been investigated by Gordon (1968).

Many skipjack tuna, after a few days in confinement, develop a condition we have termed "puffy snout" (Fig. 17). Tissues of the snout begin to swell and become edematous. The swelling spreads posteriorly until the fish is unable to close its jaws and the tissues around the eyes swell to give the appearance of sunken eyes. If left in this condition, the fish dies.

Puffy snout is believed to result from stress due to confinement in small tanks or pools. When tunas with puffy snout are placed in larger tanks or pools, the swelling recedes. We have had kawakawa which developed severe cases of puffy snout when kept in an annular raceway less than a meter wide. When these kawakawa were transferred to one of the swimming pools, the swelling disappeared within 2 weeks. Puffy snout develops more frequently in skipjack tuna than in other species, so we believe skipjack tuna are less able to adapt to confinement in space as small as our pools. Some skipjack tuna that were transported to Sea Life Park, Makapuu, Hawaii, and established in a large display tank (21.4 m top diameter, 10.7 m bottom

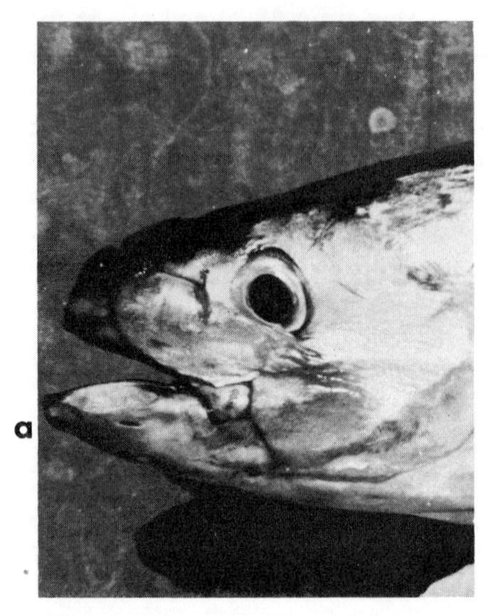

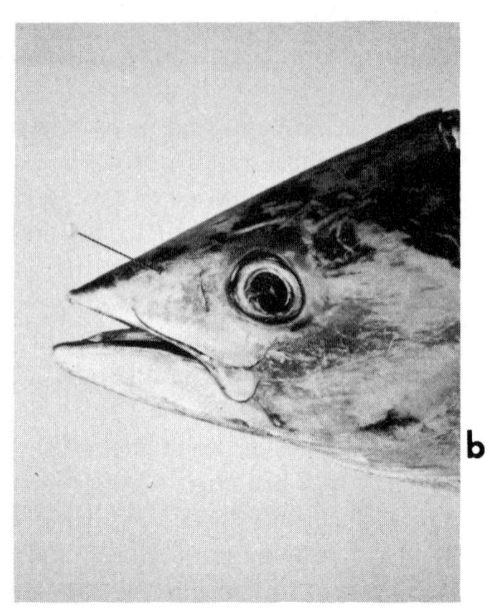

Fig. 17. Head of skipjack tuna (*Katsuwonus pelamis*). (**a**) Skipjack tuna with puffy snout. (**b**) Skipjack tuna with normal snout. Pin is stuck in anterior naris.

diameter, 4.3 m deep) did not develop puffy snout. Yellowfin tuna also have developed this malady (Tester, 1952). So have mackerels kept in small tanks. I have seen puffy snout in Pacific mackerel (*Scomber japonicus*) in an aquarium at Scripps Institution of Oceanography, La Jolla, California, and in Atlantic mackerel (*Scomber scombrus*) in an aquarium at The Plymouth Laboratory, Plymouth, England.

Larger tanks undoubtedly are more favorable for maintaining swift pelagic fishes, such as tunas, in captivity. The largest tank in which tunas have been kept in captivity is the tank for public display of fishes at Marineland of the Pacific, Palos Verdes, California. The tank is about 33 m long, 16.3 m wide, and three stories deep. Pacific bonito (*Sarda chiliensis*) have been maintained in captivity for years. These are the only tuna ever known to spawn in captivity. Courtship, spawning, swimming, feeding, and other activities of this species have been observed in this tank (Magnuson and Prescott, 1966).

In 1965–1966, Inoue *et al.* (1967) attempted to establish tunas in small pools in Japan. A pool, 1.5 m in diameter and 0.6 m deep, was used aboard a vessel at sea. Skipjack tuna died soon after placement in this pool; yellowfin tuna survived for 19 hr. A larger pool, 4 m in diameter and 0.6 m deep, was used ashore. Bonito (*Sarda orientalis*) and kawakawa were kept for over 400 hr. The authors attributed their poor success primarily to insuficient oxygen. In my judgment, the smallness of the pools was also a significant factor.

IV. CONCLUSION

Much has been learned from the use of facilities to observe tuna behavior both at sea and ashore under natural and experimental conditions. Tunas are able to detect dilute concentrations of certain chemicals (Tester *et al.*, 1955; Van Weel, 1952) and to detect sounds of certain frequencies with greatest sensitivity between 300 and 700 Hz (Iversen, 1967, 1969), but vision appears to be the most important sense used in feeding (Hsiao and Tester, 1955; Tester, 1959). Baitfish which are silvery and which are active and which take evasive action when pursued have been observed to elicit greater excitement in tunas, which makes tunas more vulnerable to the pole-and-line fishery; the greater swimming speed and the frequent changes in direction while pursuing the baitfish gives a tuna less time to discriminate between hook and bait. Visual acuity (Nakamura, 1968, 1969*b*) and swimming speed measurements (Walters, 1966; Yuen, 1966) have permitted calculations of distances of visibility of objects and the time tunas have to accept or reject a hook or evade a net (Nakamura, 1969*b*). Color patterns and color changes observed in tunas during their feeding and courtship behavior are probably used as communicating signals between fish and are certainly used by observers as signs that a fish has detected certain stimuli (Nakamura, 1962; Nakamura and Magnuson, 1965). Courtship behavior has been described for three species of tunas (Magnuson and Prescott, 1966; Hunter and Mitchell, 1967; Iversen *et al.*, 1970). Studies on feeding (by skipjack tuna) indicate that indigestible materials pass rapidly ($1\frac{1}{2}$ hr) through the alimentary canal (Nakamura, 1962), that about 15 percent of their body weight is eaten per day, that a full stomach empties in about 12 hr (Magnuson, 1969*a*), and that filter feeding (by kawakawa) is accomplished by swimming over the food rather than sucking it in (Walters, 1966). Studies of schools containing two or more species of tunas have indicated that such schools were formed by separate schools of each species drawn together by a common attractant, such as food (Yuen, 1963). Disruption of schooling orientation occurs during feeding (Nakamura, 1962). The importance of hydrodynamic and hydroacoustic stimuli in augmenting visual stimuli in schooling orientation has been noted (Cahn, 1967). Tunas without gas bladders possess negative buoyancy, and to maintain hydrostatic equilibrium they must swim continuously and use primarily their pectoral fins to obtain hydrodynamic lift (Magnuson, 1969*b*, 1970). These and other results from observations and experiments made with the use of special facilities have significantly advanced our knowledge of tuna behavior. Experience with captive tunas has indicated the ease and rapidity of conditioning them for use in experiments (Iversen, 1967, 1969; Nakamura, 1962, 1968, 1969*b*). However, the results of studies on the responses of captive tunas to various stimuli have had limited value in predicting the responses of wild tunas to the same stimuli in the open sea (Tester, 1959; Tester *et al.*, 1954).

Development of methods and facilities for studying tuna behavior is continuing. Techniques to anesthetize and revive fast-swimming pelagic fishes without injuries are being developed to permit studies requiring operations on these fishes. Larger pools are planned to maintain larger tunas and skipjack tuna in better health. New techniques and facilities for studies both at sea and ashore will continue to increase our research capabilities and will allow continued growth of our knowledge of tuna behavior.

REFERENCES

Akana, A. K., Jr., Mann, H. J., and Lee, R. E. K. D., 1960, Research vessel is fitted for underwater observation of fish, *Pacific Fisherman* **58(8)**: 8–10.

Anonymous, 1967, A vessel and sled pursue the precious "Nehu," *Commerc. Fish. Rev.* **29(7)**: 23–25.

Cahn, P. H., 1967, Some observations on the schooling of tunas (motion picture), *Am. Zoologist* **7(2)**: 64 (abst.).

Chung, S. S., Richards, E. G., and Olcott, H. S., 1967, Purification and properties of tuna myosin, *Biochemistry* **6**: 3154–3161.

Gooding, R. M., 1963, The olfactory organ of the skipjack *Katsuwonus pelamis*, Proceedings of the World Scientific Meeting on the Biology of Tunas and Related Species, July 2–14, 1962, FAO Fisheries Rep. (H. Rosa, Jr., ed.) Vol. 3, pp. 1621–1631, No. 6.

Gooding, R. M., 1964, Observations of fish from a floating observation raft at sea, Proceedings of the Hawaiian Academy of Science, 39th Annual Meeting, 1963–1964, p. 27 (abst.).

Gooding, R. M., 1965, A raft for direct subsurface observation at sea, U.S. Fish Wildlife Serv., Spec. Sci. Rep., Fish., No. 517, 5 pp.

Gooding, R. M., and Magnuson, J. J., 1967, Ecological significance of a drifting object to pelagic fishes, *Pacific Sci.* **21(4)**: 486–497.

Gordon, M. S., 1968, Oxygen consumption of red and white muscles from tuna fishes, *Science* **159**: 87–90.

Hester, F. J., 1969, Identification of biological sonar targets from body-motion Doppler shifts, *in* "Marine Bio-Acoustics" (W. N. Tavolga, ed.) Vol. 2, pp. 59–73, discussion pp. 73–74, Pergamon Press, Oxford and New York.

Hsiao, S. C., 1952, Reaction of tunas and other fishes to stimuli—1951, Part III: Observations on the reaction of tuna to artificial light, U.S. Fish Wildlife Serv., Spec. Sci. Rep., Fish., No. 91, pp. 36–58.

Hsiao, S. C., and Tester, A. L., 1955, Reaction of tuna to stimuli, 1952–53, Part II: Response of tuna to visual and visual–chemical stimuli, U.S. Fish Wildlife Serv., Spec. Sci. Rep., Fish., No. **130**, pp. 63–124.

Hunter, J. R., and Mitchell, C. T., 1967, Association of fishes with flotsam in the offshore waters of Central America, *U.S. Fish Wildlife Serv. Fish. Bull.* **66(1)**: 13–29.

Hunter, J. R., and Mitchell, C. T., 1968, Field experiments on the attraction of pelagic fish to floating objects, *J. Cons. Perma. Int. Explor. Mer* **31(3)**: 427–434.

Inoue, M., Amano, R., Iwasaki, Y., and Aoki, M., 1967, Ecology of various tunas in captivity—I. Preliminary rearing experiments, *J. Fac. Oceanogr. Tokai Univ.* **2**: 197–209.

Iversen, R. T. B., 1967, Response of yellowfin tuna (*Thunnus albacares*) to underwater sound, *in* "Marine Bio-Acoustics" (W. N. Tavolga, ed.) Vol. 2, pp. 105–119, discussion pp. 119–121. Pergamon Press, Oxford and New York.

Iversen, R. T. B., 1969, Auditory thresholds of the scombrid fish *Euthynnus affinis*, with comments on the use of sound in tuna fishing, Proceedings of the FAO Conference on Fish Behaviour in Relation to Fishing Techniques and Tactics, Oct. 19–27, 1967, FAO Fisheries Rep. (A. Ben-Tuvia and W. Dickson, eds.) Vol. 3, pp. 849–859, No. 62.

Iversen, R. T. B., Nakamura, E. L., and Gooding, R. M., 1970, Courting behavior in skipjack tuna (*Katsuwonus pelamis*), *Trans. Am. Fish. Soc.* **99(1)**: 93.

Magnuson, J. J., 1963, Tuna behavior and physiology, a review, Proceedings of the World Scientific Meeting on the Biology of Tunas and Related Species, July 2–14, 1962, FAO Fisheries Rep. (H. Rosa, Jr., ed.) Vol. 3, pp. 1057–1066, No. 6.

Magnuson, J. J., 1964*a*, Activity patterns of scombrids, Proceedings of the Hawaiian Academy of Science, 39th Annual Meeting, 1963–1964, p. 26 (abst.).

Magnuson, J. J., 1964*b*, Tuna behaviour research programme at Honolulu, *in* "Modern Fishing Gear of the World" (H. Kristjonsson, ed.) Vol. 2, pp. 560–562, Fishing News (Books) Ltd., London.

Magnuson, J. J., 1965, Tank facilities for tuna behavior studies, *Progr. Fish-Cult.* **27(4)**: 230–233.

Magnuson, J. J., 1966*a*, A comparative study of the function of continuous swimming by scombrid fishes, Abstracts of Papers Related with Fisheries, Marine and Freshwater Science, Proceedings of the 11th Pacific Science Congress, Tokyo, 1966, Vol. 7, Symposium on Biological Studies of Tunas and Sharks in the Pacific Ocean, p. 15.

Magnuson, J. J., 1966*b*, Continuous locomotion in scombrid fishes, *Am. Zoologist* **6(4)**: 503–504 (abst.).

Magnuson, J. J., 1969*a*, Digestion and food consumption by skipjack tuna (*Katsuwonus pelamis*), *Trans. Am. Fish. Soc.* **98(3)**: 379–392.

Magnuson, J. J., 1969*b*, Swimming activity of the scombrid fish *Euthynnus affinis* as related to search for food, Proceedings of the FAO Conference on Fish Behaviour in Relation to Fishing Techniques and Tactics, Oct. 19–27, 1967, FAO Fisheries Rep. (A. Ben-Tuvia and W. Dickson, eds.) Vol. 2, pp. 439–451, No. 62.

Magnuson, J. J., 1970, Hydrostatic equilibrium of *Euthynnus affinis*, a pelagic teleost without a gas bladder, *Copeia* **1970(1)**: 56–85.

Magnuson, J. J., and Prescott, J. H., 1966, Courtship, locomotion, feeding, and miscellaneous behaviour of Pacific bonito (*Sarda chiliensis*), *Anim. Behav.* **14(1)**: 54–67.

Manar, T. A., 1965, Tuna behavior—a growing field for research, *Pacific Fisherman* **63(11)**: 9–11.

Mann, H. J., 1961, Underwater observation chambers on the research vessel "Charles H. Gilbert," *in* "Research Vessel Design" (J.-O. Traung and N. Fujinami, eds.) FAO Research Vessel Forum, Tokyo, Sept. 18–30, 1961, pp. MAN 1–9.

Miyake, I., 1952, Reaction of tunas and other fishes to stimuli—1951, Part VI: Observations on sound production and response in tuna, U.S. Fish Wildlife Serv., Spec. Sci. Rep., Fish., No. 91, pp. 59–68.

Miyake, I., and Steiger, W. R., 1957, The response of tuna and other fish to electrical stimuli, U.S. Fish Wildlife Serv., Spec. Sci. Rep., Fish., No. 223, 23 pp.

Murchison, A. E., and Magnuson, J. J., 1966, Notes on the coloration and behavior of the common dolphin, *Coryphaena hippurus*, *Pacific Sci.* **20(4)**: 515–517.

Nakamura, E. L., 1960, Confinement of skipjack in a pond, Proceedings of the Hawaiian Academy of Science, 35th Annual Meeting, 1959–1960, pp. 24–25. (abst.).

Nakamura, E. L., 1962, Observations on the behavior of skipjack tuna, *Euthynnus pelamis*, in captivity, *Copeia* **1962(3)**: 499–505.

Nakamura, E. L., 1964*a* A method of measuring visual acuity of scombrids, Proceedings of the Hawaiian Academy of Science, 39th Annual Meeting, 1963–1964, pp. 26–27 (abst.).

Nakamura, E. L., 1964*b*, Salt well water facilities at the Bureau of Commercial Fisheries Biological Laboratory, Honolulu, *in* "A Collection of Papers on Sea-Water Systems for Experimental Aquariums" (J. R. Clark and R. L. Clark, eds.) U.S. Fish Wildlife Serv., Res. Rep. No. 63, pp. 169–172.

Nakamura, E. L., 1966, Fiberglass tanks for transfer of pelagic fishes, *Progr. Fish-Cult.* **28(1)**: 60–62.

Nakamura, E. L., 1968, Visual acuity of two tunas, *Katsuwonus pelamis* and *Euthynnus affinis*, *Copeia* **1968(1)**: 41–49.

Nakamura, E. L., 1969*a*, A review of field observations on tuna behavior, Proceedings of the FAO Conference on Fish Behaviour in Relation to Fishing Techniques and Tactics, Oct. 19–27, 1967, FAO Fisheries Rep. (A. Ben-Tuvia and W. Dickson, eds.) Vol. 2, pp. 59–68, No. 62.

Nakamura, E. L., 1969*b*, Visual acuity of yellowfin tuna, *Thunnus albacares*, Proceedings of the FAO Conference on Fish Behaviour in Relation to Fishing Techniques and Tactics, Oct. 19–27, 1967, FAO Fisheries Rep. (A. Ben-Tuvia and W. Dickson, eds.) Vol. 3, pp. 463–468, No. 62.

Nakamura, E. L., and Magnuson, J. J., 1965, Coloration of the scombrid fish *Euthynnus affinis* (Cantor), *Copeia* **1965(2)**: 234–235.

Nishimura, M., 1963, Investigation of tuna behavior by fish finder, Proceedings of the World Scientific Meeting on the Biology of Tunas and Related Species, July 2–14, 1962, FAO Fisheries Rep. (H. Rosa, Jr., ed.) Vol. 3, pp. 1113–1123, No. 6.

Nishimura, M., 1966, Echo-detection of tuna, *La Mer. Bull. Soc. Franco-Japon. Océanogr.* **4(3)**: 155–168.

Rayner, M. D., and Keenan, M. J., 1967, Role of red and white muscles in the swimming of the skipjack tuna, *Nature* **214(5086)**: 392–393.

Sather, B. T., and Rogers, T. A., 1967, Some inorganic constituents of the muscles and blood of the oceanic skipjack, *Katsuwonus pelamis*, *Pacific Sci.* **21(3)**: 404–413.

Strasburg, D. W., 1959, Underwater observations on the behavior of Hawaiian tuna, Proceedings of the Hawaiian Academy of Science, 34th Annual Meeting, 1958–1959, p. 21 (abst.).

Strasburg, D. W., 1961. Diving behaviour of Hawaiian skipjack tuna, *J. Cons. Perma. Int. Explor. Mer* **26(2)**: 223–229.

Strasburg, D. W., 1964, An aerating device for salt well water, *in* "A Collection of Papers on Sea-Water Systems for Experimental Aquariums" (J. R. Clark and R. L. Clark, eds.) U. S. Fish Wildlife Serv., Res. Rep. No. 63, pp. 161–167.

Strasburg, D. W., 1965, A submarine for research in fisheries and oceanography, *Trans. Joint Conf. Exhibit, MTS/ASLO, Ocean Sci. Ocean Engi.* **1**: 568–571.

Strasburg, D. W., 1966, Bureau of Commerical Fisheries operations with the submarine *Asherah* and ichthyological results of these operations, Proceedings of the Hawaiian Academy of Science, 41st Annual Meeting, 1965–1966, p. 18 (abst.)

Strasburg, D. W., and Marr, J. C., 1961, Banded color phases of two pelagic fishes, *Coryphaena hippurus* and *Katsuwonus pelamis*, *Copeia* **1961(2)**: 226–228.

Strasburg, D. W., and Yuen, H. S. H., 1960*a*, Preliminary results of underwater observations of tuna schools and practical applications of these results, Proceedings of the Indo-Pacific Fisheries Council, 8th Session, Section 3, pp. 84–89.

Strasburg, D. W., and Yuen, H. S. H., 1960*b*, Progress in observing tuna underwater at sea, *J. Cons. Perma. Int. Explor. Mer* **26(1)**: 80–93.

Strasburg, D. W., Jones, E. C., and Iversen, R. T. B., 1968, Use of a small submarine for biological and oceanographic research, *J. Cons. Perma. Int. Explor. Mer* **31(3)**: 410–426.

Suckling, E. E., 1967, Electrophysiological studies on the trunk lateral line system of various marine and freshwater teleosts, *in* "Lateral Line Detectors" (P. H. Cahn, ed.) pp. 97–103, Indiana University Press, Bloomington.

Suckling, J. A., 1967, Trunk lateral line nerves: Some anatomical aspects, *in* "Lateral Line Detectors" (P. H. Cahn, ed.) pp. 45–52, Indiana University Press, Bloomington.

Tester, A. L., 1952, Establishing tuna and other pelagic fishes in ponds and tanks, U.S. Fish Wildlife Serv., Spec. Sci. Rep., Fish., No. 71, 20 pp.

Tester, A. L., 1959, Summary of experiments on the response of tuna to stimuli, *in* "Modern Fishing Gear of the World" (H. Kristjonsson, ed.) pp. 538–542, Fishing News (Books) Ltd., London.

Tester, A. L., Yuen, H. S. H., and Takata, M., 1954, Reaction of tuna to stimuli, 1953, U.S. Fish Wildlife Serv., Spec. Sci. Rep., Fish., No. 134, 33 pp.

Tester, A. L., Van Weel, P. B., and Naughton, J. J., 1955, Reaction of tuna to stimuli,

1952–53, Part I: Response of tuna to chemical stimuli, U.S. Fish Wildlife Serv., Spec. Sci. Rep., Fish., No. 130, pp. 1–62.

Van Weel, P. B., 1952, Reaction of tunas and other fishes to stimuli—1951, Part II: Observations on the chemoreception of tuna, U.S. Fish Wildlife Serv., Spec. Sci., Rep., Fish., No. 91, pp. 8–35.

Walters, V., 1966, On the dynamics of filter-feeding by the wavyback skipjack (*Euthynnus affinis*), *Bull. Mar. Sci. Gulf Carib.* **16(2)**: 209–221.

Yoshida, H. O., 1966, Tuna fishing vessels, gear, and techniques in the Pacific Ocean, Proceedings, Governor's Conference on Central Pacific Fishery Resources (T. A. Manar, ed.) pp. 67–89, State of Hawaii.

Yuen, H. S. H., 1961, Bow wave riding of dolphins, *Science* **134(3484)**: 1011–1012.

Yuen, H. S. H., 1963, Schooling behavior within aggregations composed of yellowfin and skipjack tuna, Proceedings of the World Scientific Meeting on the Biology of Tunas and Related Species, July 2–14, 1962, FAO Fisheries Rep. (H. Rosa, Jr., ed.) Vol. 3, pp. 1419–1429, No. 6.

Yuen, H. S. H., 1966, Swimming speeds of yellowfin and skipjack tuna, *Trans. Am. Fish. Soc.* **95(2)**: 203–209.

Yuen, H. S. H., 1967, A continuous-transmission, frequency-modulated sonar for the study of pelagic fish, Proceedings of the Indo-Pacific Fisheries Council, 12th Session, Section 2, pp. 258–270.

Yuen, H. S. H., 1968, A progress report on the use of a continuous-transmission, frequency-modulated sonar in fishery research, *in* Second FAO Technical Conference on Fishery Research Craft, Seattle, May 18–24, 1968 (J.-O. Traung and L.-O. Engvall, compilers) Vol. 2, Part 4.2/VI, 17 pp.

Yuen, H. S. H., 1969, Response of skipjack tuna (*Katsuwonus pelamis*) to experimental changes in pole-and-line fishing operations, Proceedings of the FAO Conference on Fish Behaviour in Relation to Fishing Techniques and Tactics, Oct. 19–27, 1967, FAO Fisheries Rep. (A. Ben-Tuvia and W. Dickson, eds.) Vol. 3, pp. 607–618 No. 62.

Chapter 7

REGULATION OF FEEDING BEHAVIOR OF THE BICOLOR DAMSELFISH (*EUPOMACENTRUS PARTITUS* POEY) BY ENVIRONMENTAL FACTORS

Robert A. Stevenson, Jr.

Rosenstiel School of Marine and Atmospheric Science
University of Miami
Miami, Florida

I. INTRODUCTION

Studies of food chains and webs on reefs have been carried out by collecting and observing animals, identifying and quantifying the kinds of food they have eaten, and thereby determining their ecological relationships. Emery (1968), Hobson (1968), Randall (1967), Hiatt and Strasburg (1960), and others showed that certain species of fishes feed predominantly on particular kinds of animals or plants and thus have established pathways through which food, and energy, "flows."

Large amounts of energy exist in the form of great numbers of fishes in reef ecosystems. The energy contained in this biomass originates from various sources. Many fishes such as some species of damselfishes (Pomacentridae), wrasses (Labridae), and cardinal fishes (Apogonidae) bring energy into the system from an outside source in the form of plant and animal plankton upon which they feed. Species of surgeon fishes (Acanthuridae) and parrot fishes (Scaridae) primarily cycle energy that is already present in the system by feeding on benthic algae or corals. Other fishes such as drums (Scaenidae) and goat fishes (Mullidae) feed carnivorously on crustaceans and mollusks that also are already a part of the ecosystem; piscivorous species such as scorpion fishes (Scorpaenidae) and groupers (Serranidae) cycle energy within the system since they feed on fishes in reef communities.

The type of feeding behavior of these reef fishes determines how energy is passed from one organism to the next. Feeding behavior has been determined through direct observations as well as by identifying stomach contents. It is known, for instance, that species of goat fishes and drums are basically benthic feeders, whereas grunts (Pomadasyidae) are benthic as well as midwater feeders and some damselfishes feed almost exclusively above the bottom on plankton (Hobson, 1965; Gosline and Brock, 1960; Suyehiro, 1942). Fishes obtain their food by picking prey from the water, "feeling" along the bottom with specialized sensory apparatus, blowing away or scraping substrate, luring other species within range and seizing them, or by other means. Stalking by camouflaged predators such as scorpion fishes (Scorpaenidae) and rushing by midwater species such as jacks (Carangidae) are behavioral adaptations enabling these species to catch more elusive kinds of prey (Wickler, 1967; Abel, 1962; Breder, 1946). Such studies enable us to understand the role of behavioral adaptations in the energetics and energy acquisition within the reef ecosystem.

There is, however, a fundamental void in this picture of behavioral ecological relationships. Seldom shown in studies is the way in which the environment regulates energy inflow through its influence on the behavior of fishes. We must start here if we are to further understand important factors that regulate inflow and cycling of energy in the animal fraction of reef ecosystems.

Since feeding behavior is regulated by environmental influences, it is of fundamental importance to study the animal's behavior in relation to these influences. One of the most basic environmental influences is light. Its effects on feeding, spawning, and other activities of fishes have been summarized by Woodhead (1966). Recently, Hobson (1965, 1968) has made extensive studies by directly observing shore fishes in the Gulf of California both by day and night. His observations clearly show how different species react to light by feeding at discrete times during a 24-hr period. Most species confined their feeding either to daylight periods, to night-time, or to the interim crepuscular periods. Schools broke up during the evening crepuscular period, and individual fish spread out across the bottom to feed on organisms that became active at night. Piscivorous species of fishes also became active during this period, and predation sharply increased. Hobson's work gives us a good background of how light regulates the acquisition and cycling of energy by regulating behavioral activity. Knowing the times of day during which many species feed provides us with a basis from which to make more quantitative studies of relationships between light and feeding behavior and with which to compare other environmental factors that have effects on feeding behavior.

The author has attempted to explore such avenues of research by studying the damselfish group, which represents a large biomass on many coral reefs. Some species may crop algae or pick small organisms from the bottom; other swim above it, where they feed on plankton; others do both. Those species that feed in the water column provide a pathway that brings energy into the ecosystem from an open component, whereas those that feed on the bottom mostly recycle energy that the system already contains. A certain amount of recycling occurs when planktonic stages of benthic organisms are eaten by the species of damselfishes that feed in the water column.

Some species of damselfishes within the genera *Eupomacentrus*, *Chromis*, and especially *Dascyllus* have behavioral patterns that are logistically suited for detailed quantitative studies. Their feeding behavior can be observed throughout the daylight hours as they catch planktonic organisms above the bottom. Reproductive activities are also observable because these "nest-building" species carry out their activities around a particular spot on the bottom. This localization of individuals is not restricted to reproductive males because immature individuals and mature females and males also remain in one spot for months at a time. Groups of damselfishes characteristically cluster around some prominent object such as a rock or coral head which they use as shelter.

The damselfish searching activity culminates with a characteristic grabbing motion or "deflection," as we have called it, that occurs when they inspect or engulf food. Another important activity that is related to feeding behavior also can readily be seen. The distance that an individual fish maintains between itself and the sheltering rock or coral head where it hides from view when threatened is significant. This distance can be considered a behavioral process influenced by external environmental factors. Essentially, the distance maintained is probably a balance between the tendency to swim toward current to get food and the countering tendency to return to the protection of shelter.

From observations of damselfishes and their positioning around shelters, it is apparent that both the rates of feeding and the positions, or spatial relationships, change. In some cases, it is possible to determine the environmental influences that bring about some of these changes. The positions of juvenile *Dascyllus albisella*, for instance, change when they are in shallow water and surge is present (Stevenson, 1963). Individual passing surges physically push the aggregation of fish back and forth, toward and away, from coral heads. Occasional shifts in position also result wherever these fish congregate when individual fish dart after food particles. In the evening and morning when light levels are low, and during times when predators threaten, the aggregations remain closer to shelter than at other times, which indicates that these spatial relationships vary with potential predators and with light intensity. This pattern of activity is general among many pomacentrids and is familiar to observers who have spent long periods of time underwater observing fish behavior.

Fishes that form these feeding aggregations spend most of the daylight hours feeding on plankton, their rates varying with shifts in positioning. When darkness is near, or when predators threaten, they remain close to shelter and appear to feed little. Conversely, when they leave shelter, their feeding rates immediately appear to increase. Proximity to shelter, then, seems to be correlated in some way with decreased feeding activity. Since factors such as light, surge, and distance from shelter seem to affect feeding, such factors are related to amounts of energy accumulated in the form of fish biomass.

The author studied the plankton-feeding bicolor damselfish (*Eupomacentrus partitus*) at the Lerner Marine Laboratory, Bimini, Bahamas, from 1967 to 1968. These studies revealed some of the ways in which behavior is affected by environmental conditions to create fluctuations in feeding behavior. The bicolor damselfish is a reef-inhabiting species with behavior typical of that already cited. Aggregations of all sizes of individuals ranging from 15 mm juveniles to 75 mm adults gather around coral rocks and sometimes sponges, where they feed on plankton that drifts past them. Adult males acquire territories and prepare "nest" sites to which they attract feeding females from their own or nearby aggregations. Like many other fishes, the youngest juveniles remain close to sheltering rocks and succeedingly larger ones swim greater distances up to about 1.5 m into the water column after food. They feed during daylight hours and retire to cover during the evening and remain there until light returns in the morning.

II. METHODS

The use of underwater television permitted continuous study of fish throughout the day and also allowed the use of recording instruments that could not be used underwater. The use of videotape made it possible to observe behavioral and environmental conditions more-or-less simultaneously by rerunning the tapes. Arbitrary zones of water around a rock that sheltered an aggregation of eight bicolors were established to judge the distances of known individuals from cover. One zone was considered as "on the rock," another extended from the surface of the rock to a distance of about 30 cm, another from 30 to 60 cm and the farthest extended from about 60 cm to about a meter away from the rock (Fig. 1). Currents flowed at right angles to the camera, and fish generally swam directly into it. How-

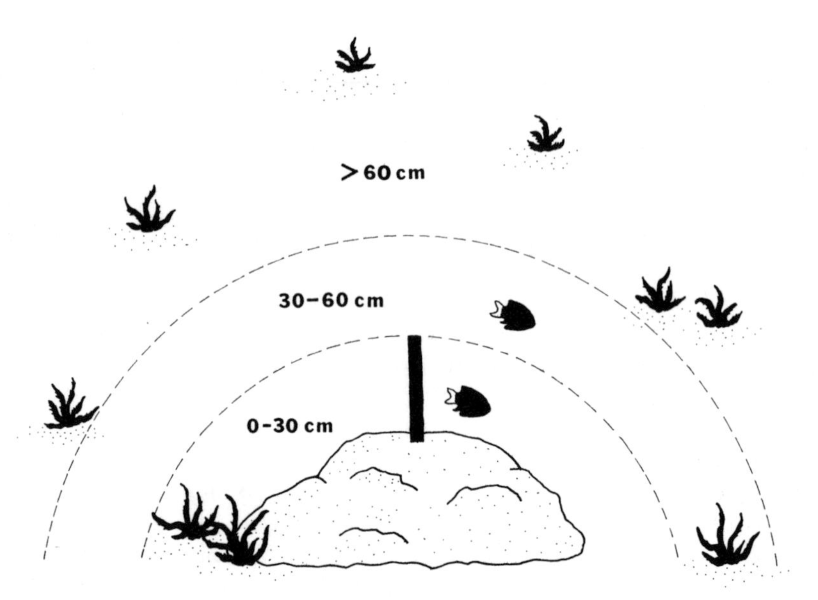

Fig. 1. Diagram of study area showing zones used to judge positions of fish around sheltering rock.

ever, fish occasionally swam away or toward the camera, which introduced some bias into distance measurements.

Each fish usually was observed for 9 min during the morning, around midday, and later in the afternoon or evening for a total of about 27 min each day, and the time it spent in each zone was recorded. Changes in light intensities and current speeds could be observed by watching instruments that indicated these characteristics of the environment at the 20 m depth where the work was conducted. Feeding rates were determined by counting the number of "deflecting" motions. Because the aggregation was small, it was possible to recognize a particular dominant male, a mature female that mated with it, and a juvenile. The author could follow their behavior separately as well as together using the videotape recorder. Observations were made on these fish during the summer of 1967 and on different individuals at the same location during the winter of 1968.

The figures presented here were constructed in a way to show graphically where a particular fish spent most of its time in the water column. Thus, if the widest part of a histogram appears at the top, then a fish spent the greatest proportion of a 3-min period farthest from the rock in the >60 cm zone. If the widest portion of the histogram appears at the bottom in a figure, then the fish spent most of a 3-min observation period on the rock shelter. In scanning the histograms it is less confusing if one watches the lowest portion or "on rock" zone, as this represents where the fish is not feeding whereas zones above this are where feeding occurs almost exclusively.

III. RESULTS

A. Distance from Rock

It was found in the summer that when current was flowing (0.2–0.4 knots) during well-lighted portions of the day (>350 ft-candles), an adult and juvenile spent little time on the rock during the middle and late portions of the day (Fig. 2 and 3). On days when current was not flowing, they tended to spend more time on the rock in the afternoon than when it was flowing. The fish swam vigorously into the current, often crossing back and forth upstream from the rock as they fed on planktonic organisms. This pattern was repeated in the winter by a different adult and juvenile (Fig. 4 and 5).

During the well-lighted daylight hours, behavior further varied with the presence or absence of current as follows: If current was not flowing in the morning, both adults and older juveniles ranged beyond 60 cm from the rock during those periods when they were in the water column. If current was flowing, adults still tended to achieve these distances, but juveniles characteristically remained between the adults and the rock. During the afternoon when current ceased to flow, both adults and juveniles swam far

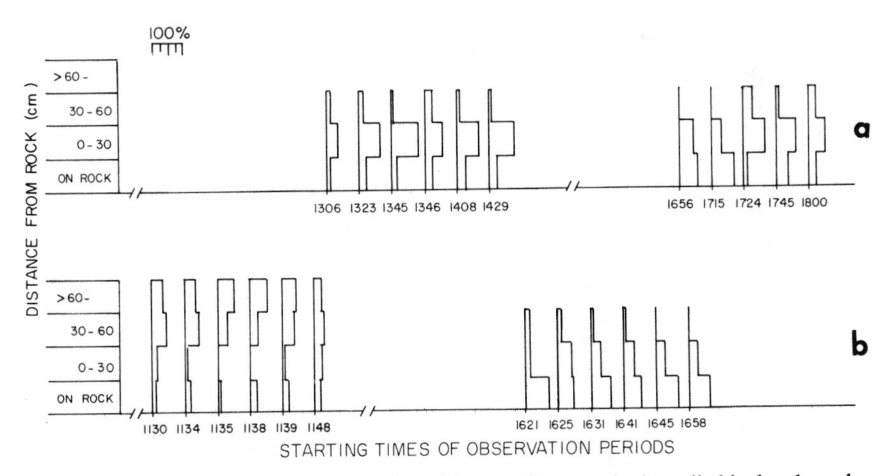

Fig. 2. Relation between current and time of day to distance of a juvenile bicolor damselfish from shelter in the summer of 1967. Current was **(a)** present 3 days (0.2–0.4 knots) and **(b)** absent 3 days under conditions of "full" daylight (>350 ft-candles). Each histogram (scale at upper left) shows the percentage of a 3-min observation period spent at a particular distance.

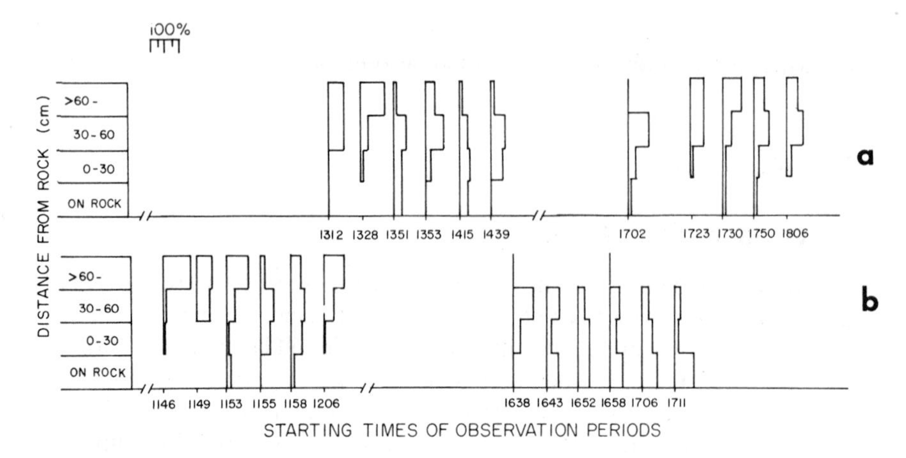

Fig. 3. Relation between current and time of day to distance of an adult female bicolor damselfish from shelter in the summer of 1967. Current was **(a)** present 3 days (0.1–0.3 knots) and **(b)** absent 3 days under conditions of "full" daylight (>350 ft-candles). Each histogram (scale at upper left) shows the percentage of a 3-min period spent at a particular distance.

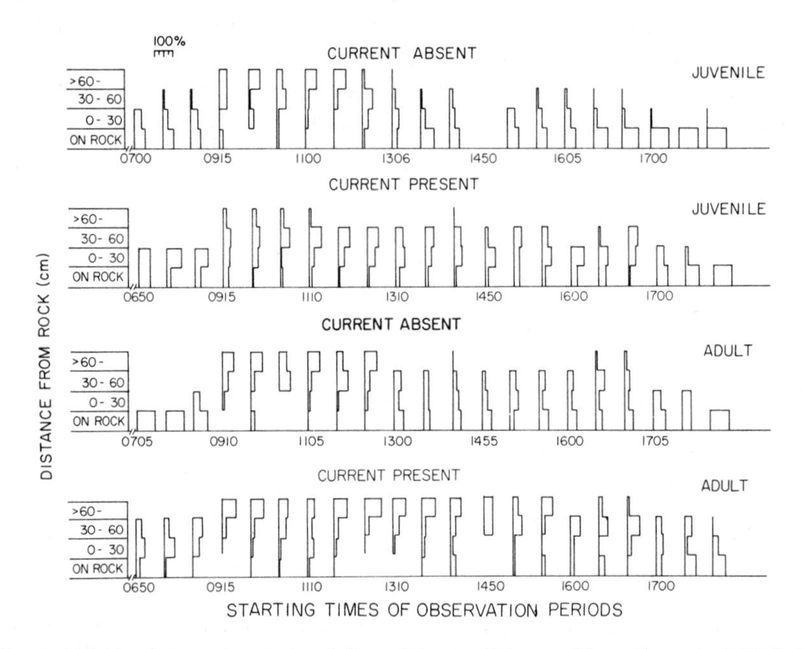

Fig. 4. Relation between current and time of day to distance of juvenile and adult bicolor damselfish from shelter in the winter of 1968. Each histogram (scale at upper left) shows the percentage of a 3-min observation period spent at a particular distance. Light intensities ranged from 20 to more than 350 ft-candles and current from 0 to 0.3 knots.

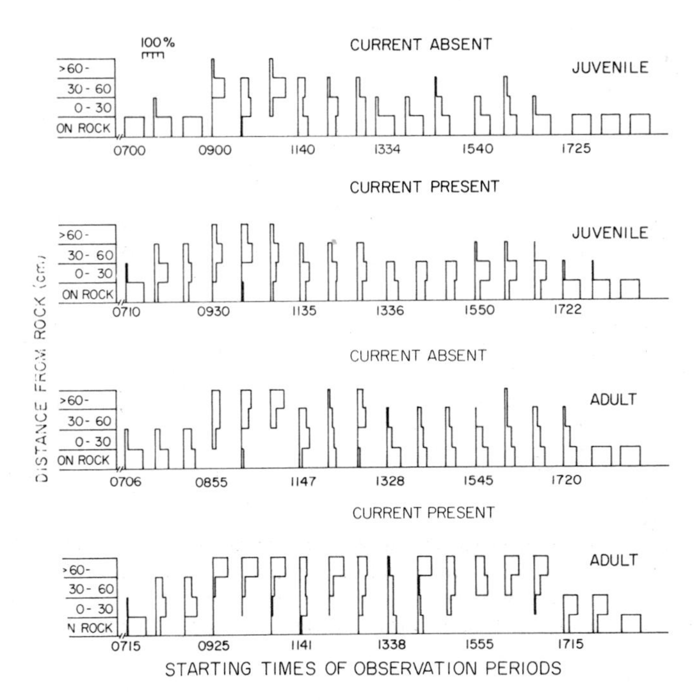

Fig. 5. Relation between current and time of day to distance of juvenile and adult bicolor damselfish from shelter in the winter of 1968. Each histogram (scale at upper left) shows the percentage of a 3-min observation period spent at a particular distance. Light intensities ranged from 15 to more than 350 ft-candles and current from 0 to 0.2 knots.

less vigorously than in the morning when current was not flowing, and both remained closer to the rock than on afternoons when current was flowing. The adults almost continuously remained farther from the rock than juveniles whenever fish were in the water column during the day. An adult male guarding its territory ranged throughout the water column and consistently returned to tend an area of rock that it kept clean of growth (Fig. 6). As a whole, this fish seemed to make more regular excursions to all levels of the water column than the adult female.

The behavior described thus far is that which occurs during hours of full daylight (> 350 ft-candles). As we have seen, the changes in spatial relationships that occurred were influenced largely by the presence or absence of currents. Spot checks during the daytime confirmed that such behavior remained quite constant over many hours. During crepuscular periods, however, large changes in spatial relationships and feeding rates occurred over a span of minutes.

During the evening crepuscular periods, both feeding behavior and distance relationships diminished rapidly, while during the morning they

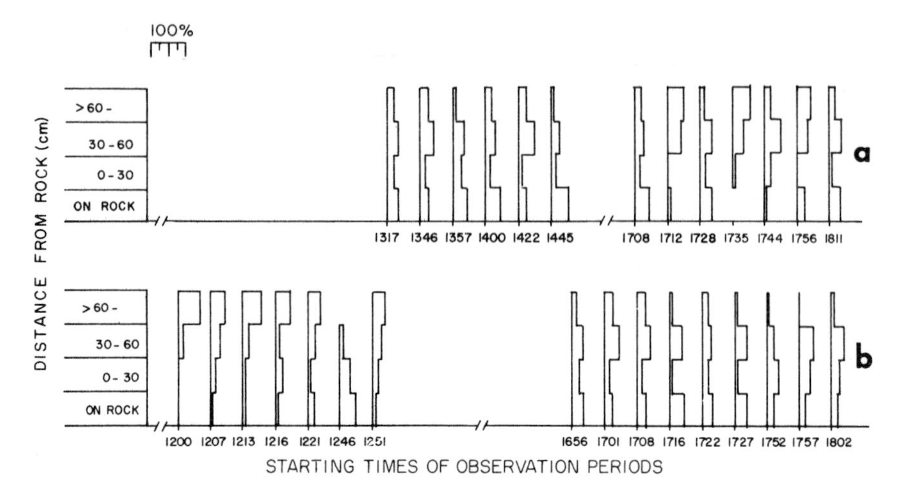

Fig. 6. Relation between current and time of day to distance of an adult male bicolor damselfish from shelter in the summer of 1967. Current was **(a)** present 3 days (0.2–0.4 knots) and **(b)** absent 3 days under conditions of "full" daylight (>350 ft-candles). Each histogram (scale at upper left) shows the percentage of a 3-min period spent at a particular distance.

increased rapidly from zero to full feeding and development of longer distance relationships. Whereas the primary environmental stimulus for feeding during full daylight hours is current, during the crepuscular periods the outstanding stimulus is a changing light condition.

It can be seen from Fig. 7(b) that both juvenile and adult bicolors remained almost entirely on the rock at low light levels during evening crepuscular periods when current was absent. However, on evenings when current was flowing at 0.1 or more knots during comparable light levels, only the juvenile and adult female fishes remained away from shelter for considerable periods (Fig. 7a). Thus, in the presence of current at low light levels the tendency to leave shelter and search for food prevailed over the tendency to remain close to shelter. As before, adult fishes moved farther from shelter than the juveniles, showing their greater independence. This pattern was repeated by a different adult and juvenile during the following winter (Fig. 8), which indicates that seasonal differences in this behavior do not exist.

Figure 9 compares the same juvenile and adult female at comparable light levels during morning (a) and evening (b) crepuscular periods. It is apparent that the light levels at which bicolors leave shelter in the morning crepuscular period are lower than the levels at which they return to shelter in the evening.

The presence of current in the morning crepuscular period also resulted in fish leaving shelter at lower levels than when it was absent (Fig. 10). These winter data, based on different fish from those observed in the summer, also show the tendency of adults to leave shelter sooner and to go farther away than juveniles. Individual differences may have contributed to fish leaving shelter at somewhat higher light intensities than in summer, although seasonal effects and perhaps other environmental factors might have been involved.

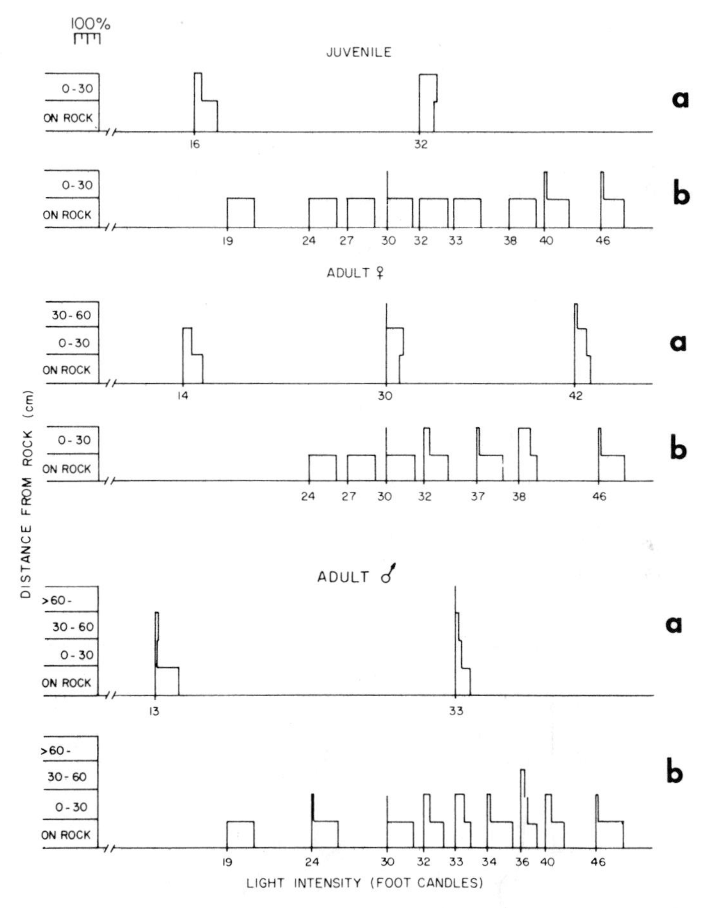

Fig. 7. Relation between current and distance from shelter of juvenile and adult bicolor damselfish at low light intensities during evening crepuscular periods in the summer of 1967. **(a)** The juvenile and adult male were observed 1 day and the adult female 2 days when current was flowing (0.1–0.2 knots). **(b)** All fish were observed 3 days when current was absent. Each histogram (scale at upper left) shows the percentage of a 3-min observation period spent at a particular distance.

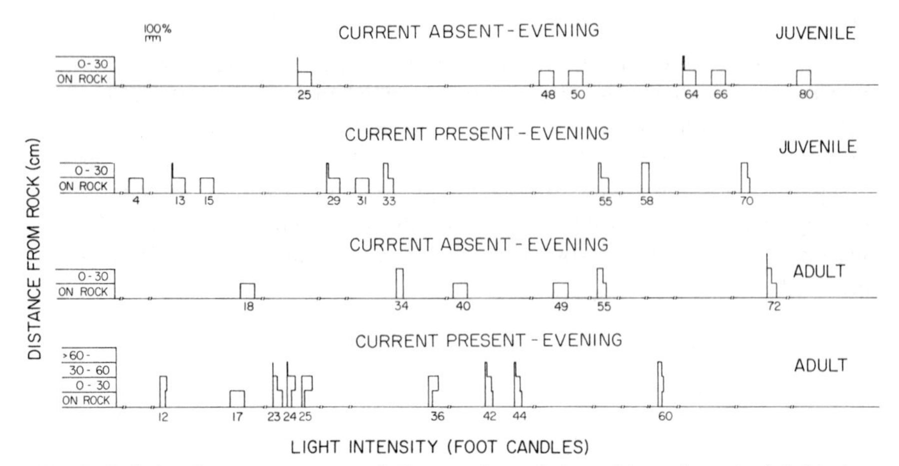

Fig. 8. Relation between current and distance from shelter of juvenile and adult bicolor damselfish at low light intensities during evening crepuscular periods (current speed 0.1–0.15 knots) in the summer of 1967. Each histogram (scale at upper left) shows the percentage of a 3-min observation period spent at a particular distance.

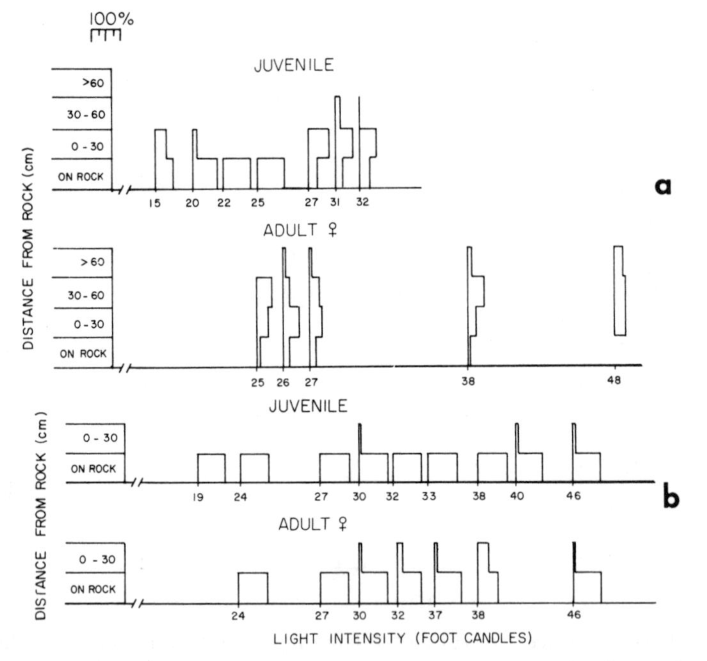

Fig. 9. Comparison of distance from shelter of an adult female and a juvenile bicolor damselfish at low light intensities during morning crepuscular periods on 3 days when current was (**a**) absent in the summer of 1967. These data may be compared with (**b**) (taken from Fig. 5), which shows the behavior at comparable light intensities during evening crepuscular periods. Each histogram (scale at upper left) shows the percentage of a 3-min observation period spent at a particular distance.

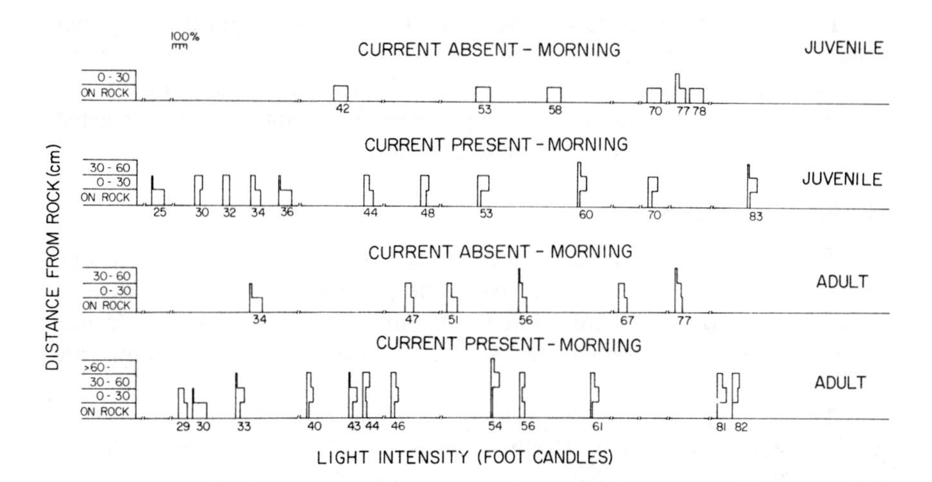

Fit. 10. Relation between current (0.05–0.15 knots) and distance from shelter of juvenile and adult bicolor damselfish at low light intensities during morning crepuscular periods in the summer of 1967. Each histogram (scale at upper left) shows the percentage of a 3-min observation period spent at a particular distance.

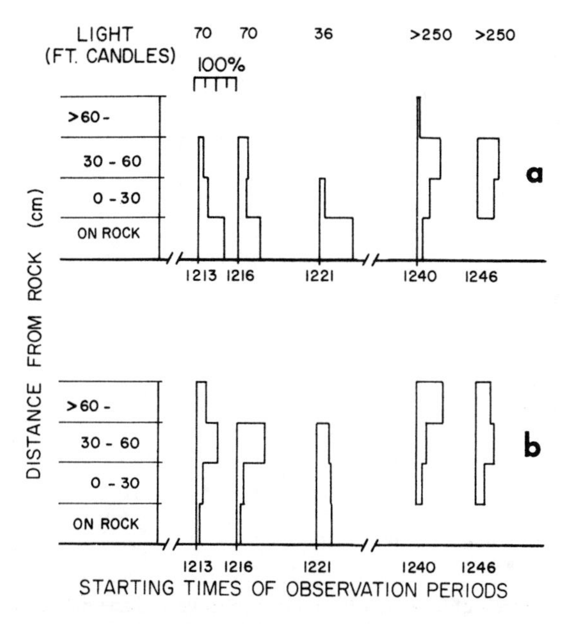

Fig. 11. Effect of changing light intensities due to clouds on distance from shelter of (**a**) a juvenile and (**b**) an adult female bicolor damselfish in the absence of current in the summer of 1967. Each histogram (scale at upper left) shows the percentage of a 3-min observation period spent at a particular distance.

It is important to note that fish react to changes in light intensity that occur at times other than during the crepuscular periods. Figure 11 shows how two bicolors moved closer to shelter with passage of heavy cloud cover and again moved away as light levels increased. The adult fish, again, reacted less strongly than the juvenile to these fluctuations.

B. Feeding Rates

The effects of current on the feeding rate of bicolors can be seen in Fig. 12 and 13. During full daylight hours, the presence of current resulted in increased feeding in both the adult female and juvenile bicolors. Under the same conditions, the adult male showed no clear increase in feeding rate (Fig. 14 and 15).

The presence of current prolonged feeding time during evening crepuscular periods (Fig. 16). The juvenile and adult female observed in the summer of 1967 actively fed at low light intensities if current was present. Feeding was essentially nil when current was absent at low light intensities. The presence of current during morning crepuscular periods also promoted

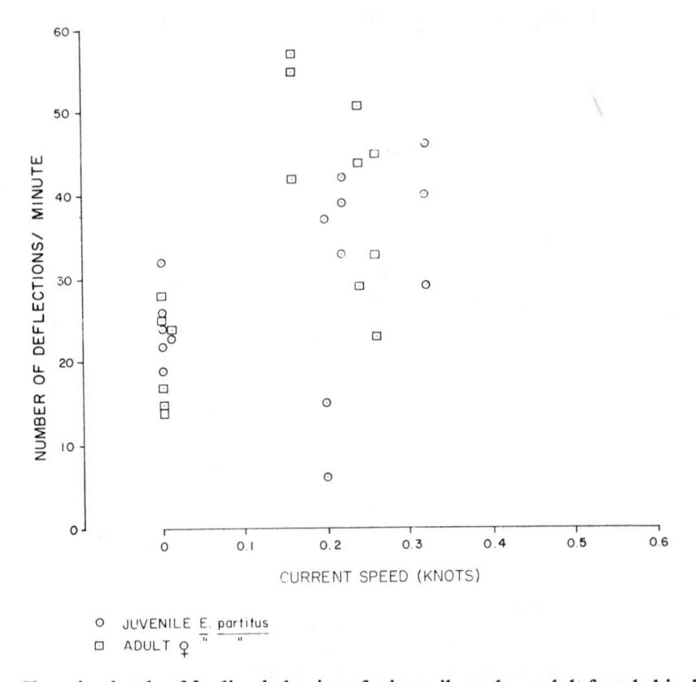

Fig. 12. Changing levels of feeding behavior of a juvenile and an adult female bicolor damselfish with different current speeds. Data were compiled over 6 days in the summer of 1967. Each point is based on a 3-min observation period between 1130 and 1439 hr when light intensities were >350 ft-candles.

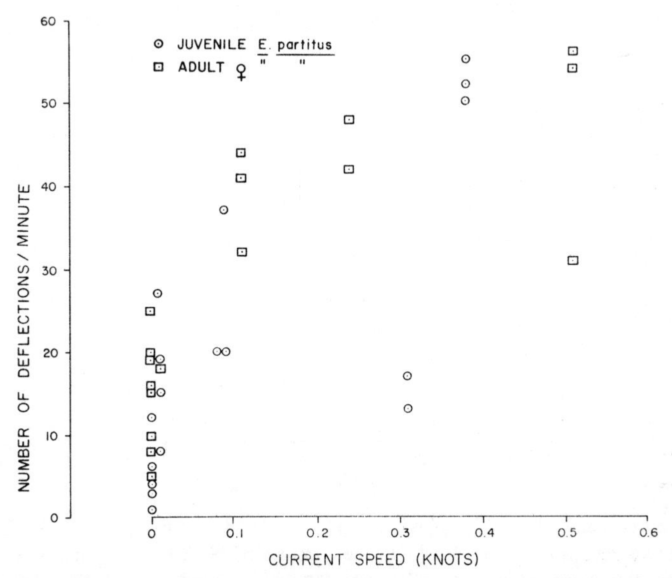

Fig. 13. Changing levels of feeding behavior of a juvenile and an adult female bicolor damselfish with different current speeds. Data were compiled over 6 days in the summer of 1967. Each point is based on a 3-min observation period between 1621 and 1806 hr when light intensities were >350 ft-candles.

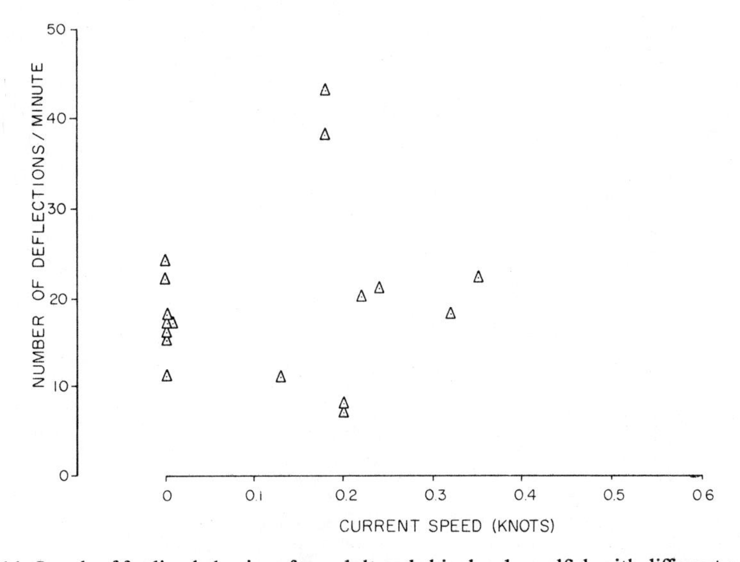

Fig. 14. Levels of feeding behavior of an adult male bicolor damselfish with different current speeds. Data were compiled over 6 days in the summer of 1967. Each point is based on a 3-min observation period between 1130 and 1439 hr when light intensities were >350 ft-candles.

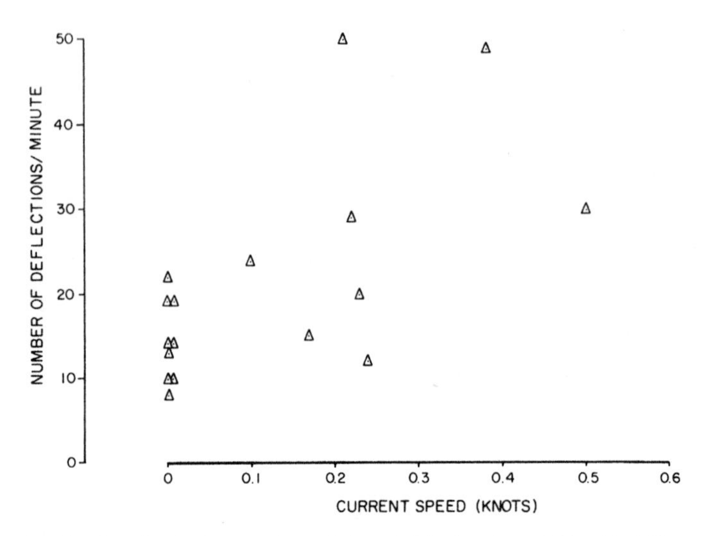

Fig. 15. Levels of feeding behavior of an adult male bicolor damselfish with different current speeds. Data were compiled over 6 days in the summer of 1967. Each point is based on a 3-min observation period between 1621 and 1806 hr when light intensities were >350 ft-candles.

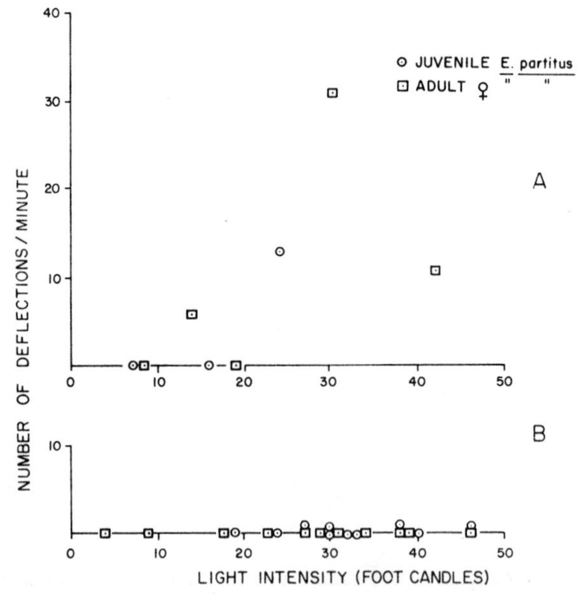

Fig. 16. Relation between levels of feeding behavior of a juvenile and an adult female bicolor damselfish and currents during evening crepuscular periods. Current (0.1–0.2 knots) was (**a**) present on 2 evenings and (**b**) absent on 3 in the summer of 1967). Each point is based on a 3-min observation period.

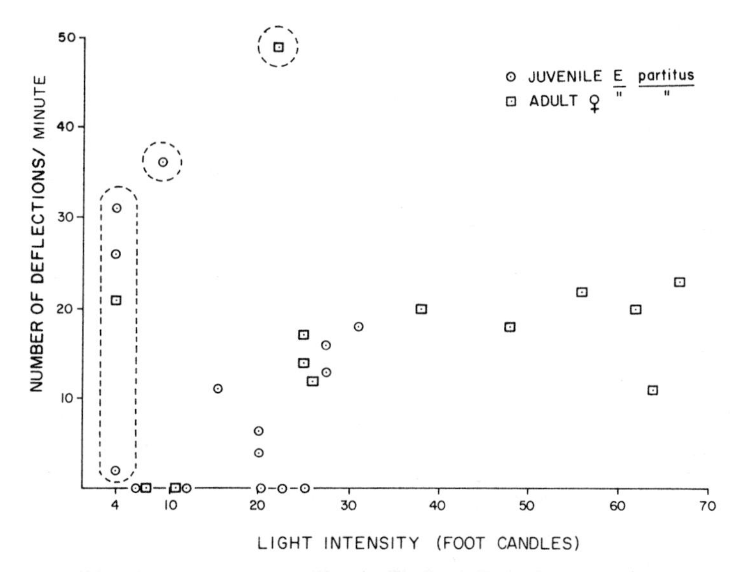

Fig. 17. Relation between levels of feeding behavior of a juvenile and an adult female bicolor damselfish and current during morning crepuscular periods. Current (0.1–0.3 knots) was present 2 mornings (dashed portions) and absent 4 mornings (in the summer of 1967). Each point is based on a 3-min observation period.

feeding at lower light intensities, which again is a reflection of fishes leaving shelter (Fig. 17).

The effect of lowered light intensity during midday periods is similar to that found during crepuscular periods (Fig. 18). On one day when current was present with low light intensities around midday, feeding rates were higher than on another day when current was absent. Distance relationships again are involved as the fish interact with currents and light.

Figure 19 brings together data on feeding rates from 2 days when current was absent and 2 days when it was present during the summer of 1967. The total number of feeding deflections performed by the male, female, and juvenile were averaged for the three 3-min periods during each time of day sampled and single standard deviations were calculatted. Although the deviations are large, it is apparent that the presence of current results in much higher rates of feeding.

IV. DISCUSSION

The food income of bicolor damselfish is strongly linked to spatial relationships that result from two drives or tendencies regulated by currents

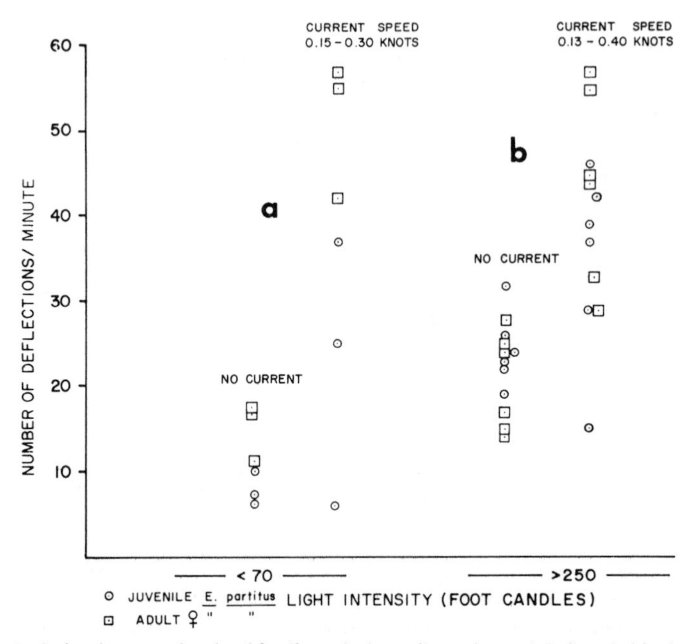

Fig. 18. Relation between levels of feeding of a juvenile and an adult female bicolor damselfish and current when light levels were variable around midday. Current was (**a**) present 1 day and absent 1 day when light was low and (**b**) absent on 2 days and present on 2 days when light was high (in the summer of 1967).

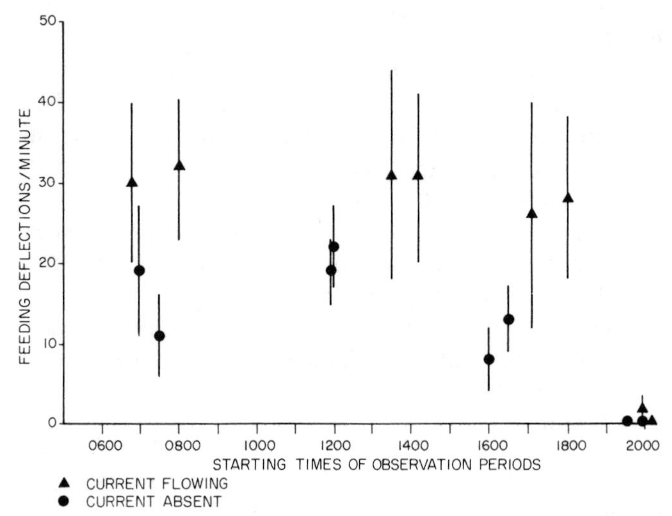

Fig. 19. Average feeding rate of three bicolor damselfish on 2 days when current was absent and on 2 days when it was present in the summer of 1967. Light intensities ranged from 5 to more than 350 ft-candles.

and light. These drives are (1) the tendency to remain close to shelter and (2) the tendency to leave shelter to feed. During the morning hours after full daylight prevails, feeding activity is vigorous. At this time, the tendency to feed dominates the tendency to remain close to shelter and fish swim farther from their shelter in search of food. If there is no current to bring plankton, the fish exhaust the food supply near the shelter and then move maximum distances to find plankton. Later in the morning and especially in the afternoon when current is absent, lack of reinforcement through the fruitless search for food occurs and the tendency to seek shelter prevails. On the other hand, the presence of current and therefore plankton strengthens the tendency to feed, and fish remain away from shelter even during the afternoon. The tendency to return to shelter is less pronounced in adults, which often remain in the water column after juveniles have returned to shelter. The conflict between remaining close to and leaving shelter is general, and occurs with many benthic fishes, and results in a gradation of sizes away from shelter with small fishes remaining close to and larger fishes farther away from shelter.

In summary, the least complicated situation occurred during days when light regulated feeding in the absence of current. During morning crepuscular periods, feeding began over a period of about 15–20 min from the time that fish began to make occasional sorties beyond the confines of their rock shelter. Maximum feeding levels for these conditions were established within about 45 min and were held until midday. Feeding levels then began to fall and decreased to approximately half the maximum rate by midafternoon and to almost nil by about $\frac{1}{2}$ hr before the evening crepuscular period.

The amount of food inflow under these conditions changes when current is present. In the morning, fish reach near peak feeding within about 5–10 min from the time they leave shelter. These levels of feeding are almost double those that occur when no current is present at comparable light levels in the morning. They continue into the afternoon for some 4–5 hr after they would have decreased if no current were present. Lower levels of feeding persist almost to darkness, giving about one half hour of additional feeding time compared to noncurrent situations. The major significance in relation to inflow of food is that these higher rates extend over a number of hours during afternoons and crepuscular periods when rates fall if current is absent.

An occasional fortuitous observation helped to show the extent to which light regulates the behavior of bicolor damselfish. Throughout one summer morning, feeding activity was slow and fish remained close to their rock shelter. The only difference between that morning and the one the day before when fish fed actively was the presence of heavy cloud cover. When considered with other evidence presented herein, it seems that greatly reduced light intensity can reduce food inflow by depressing activity of bicolor damselfish at almost any time of day. However, the presence of current and heightened feeding drives due to food deprivation probably stimulate feeding to some extent during cloudy periods.

It is also possible that clouds might stimulate feeding under certain conditions. On three occasions, cloud banks on the eastern horizon strongly reflected light into the water for 10–15 min after light normally fails rapidly at sunset. With favorable current situations, fish might extend their period of feeding during such times, thus increasing the amount of food inflow.

Because currents transport food to benthic plankton-feeding fishes, food inflow is affected by tidal flow in areas where such currents are found. In some areas, wind-induced currents also must be taken into consideration. The author has observed that wind-induced currents intensify feeding behavior at Bimini, and it is probable they may prolong feeding when other currents are absent. Close attention to tidal and meteorologically induced currents, therefore, is of considerable importance when evaluating food income in fishes that depend on such conditions.

The sex and stage of reproductive readiness of the bicolor damselfish influence the time that it spends feeding. The reproductive male spends much time chasing intruders, picking at the substrate, and in other activities. When eggs are in the "nest," the male's behavior is directed even more toward territorial and parental activities and less toward feeding. Mature males without territories and mature females spend their time feeding, except for relatively brief periods when the latter may be on the nest spawning. In one case, spawning which usually occurs very early in the morning was observed to occur for a period of about 40 min. Such periods may be relatively infrequent, however, and only a small percentage of fish are involved in this activity at any particular time.

Individual differences in feeding rates are another source of variability in food income. Similarly sized bicolors, on occasion, have been observed feeding at noticeably different rates while swimming side by side. If individuals feed continuously at different rates, then calculations of intake would have to take this into consideration. The quantity and quality of food available can be expected to influence feeding rate because of factors such as selectivity in feeding among individuals and degree of satiation. The external environment also may contribute to individual differences. On one occasion, it was noted that a bicolor was feeding at a much higher rate than others but also was swimming unusually high above the bottom. Subsequent diving operations revealed that current close to the bottom could be essentially zero while a considerable flow of water could exist less than a meter above. It was apparent that by ranging farther from shelter the fish was able to obtain more food than others that exhausted the supply of plankton in the slowly moving waters near the bottom.

Feeding rates and feeding time are also affected by turbidity. It was observed that when visibility decreased from 15–20 m to 3–4 m all bicolors withdrew from the water column and remained close to the sheltering rock. Because of the hydrography of the area, masses of clear water occasionally were interspersed with turbid water. Fish reacted quickly and consistently to the alternating periods of unobstructed and obstructed vision by either leaving or returning to shelter, respectively. On days when heavy turbidity was present for hours at a time, fish were always close to shelter during those periods when they were visible to the observer. Their feeding rates also were clearly impaired by turbidity. Emery (personal communication) made similar observations on a number of pomacentrid fishes and noted that after a period of several days of turbid water they began to range farther from shelter in spite of turbidity. Interactions such as these which are probably due to extreme food deprivation are important considerations in food acquisition and show the important role of hunger drives in feeding relationships in the field.

The important question of predation on this species, which brings energy acquired by these plankton-feeders into the other cycles in the ecosystem, can be only fitted loosely into the basic framework of their almost continual feeding behavior. In some 200–300 hr of observation during daylight hours, no obvious attack upon a bicolor damselfish was seen. The potential existed, however, when groupers occasionally visited the shelter and frequently, yellow tail snappers (*Ocyurus chrysurus*) would invade the area and spend a few minutes apparently examining the shelter and occasionally striking down at it. Many observations on pomacentrids in the Hawaiian Islands by the author also have failed to reveal actual attacks on feeding fishes during the daytime, although aggregations frequently retreated to shelter in alarm from unseen causes.

Infrequent attacks during the daytime by free-swimming predators is probably related to the behavior pattern of pomacentrids such as the bicolor damselfish which remain close to shelter. The difficulties that free-ranging predators such as jacks (Carangidae) and barracudas (Sphyraenidae) have in catching fishes that live close to the bottom is understandable considering their proximity to shelter as well as the community reaction which sends many fishes into hiding. Most predation probably comes from bottom-living predators, which are in a favorable position to exploit plankton-feeding and other species when they are close to the bottom, especially during the crepuscular periods. Hobson (1968) remarked that "most diurnal predators feeding largely on fishes stalk their prey among the rocks . . . or rush suddenly from cover to capture prey that have ventured near." Emery (1968) observed predation on the bicolor damselfish on many occasions and stated that "the most common situation on the reef top was during reproductive signal jumping." Emery (personal communication) also stated that this species was somewhat more vulnerable to predation when they ranged farther from shelter during long periods of turbid water.

One factor having potentially wide importance in food income involves community makeup within ecosystems. It is apparent by now that many fishes return to the same locations after feeding elsewhere during diurnal or nocturnal periods (Hobson, 1968; Cummings, 1968; Stevenson, 1963; Randall, 1961; Bardach, 1958). Fishes forage to varying distances that range from centimeters to hundreds of meters. These behavioral characteristics, which have a profound effect on food habits, prompted Randall (1961) to collect in all marine environments in the Virgin Islands and Puerto Rico and sample fishes for stomach contents. He stated that "food habits may vary markedly with the environment from which the fish are collected." The effect of diverse reef habitats on food habits of fishes is further illustrated by Longley *et al.* (1925). They stated that the stomach contents of gray snappers (*Lutjanus griseus*) were so different that it would be possible to determine from an average sample of ten fish from which of seven different sites the fish had been taken. They felt that individuals only moved limited distances from areas where they congregated during the day. They also pointed out the need to study the microgeographic distribution of more sedentary food organisms in relation to distances that fishes range for food. Restricted movements, therefore, result in particular individuals remaining in localities on reefs where the supply of food may be of different quality and quantity than at other locations. Benthic plankton-feeders also are nonrandomly distributed, as are the planktonic organisms on which they feed. The nonrandom distribution of both fishes and their food that leads to nonrandom intake of food is an important factor to remember when considering uptake and cycling of energy. Certainly, knowledge of such details of ecosystem dynamics would aid in accounting for variability encountered in actual measurements of energy.

The relationships between light, currents, and behavior presented here likely apply to many plankton-feeding pomacentrids. Many authors have referred to feeding formations of *Eupomacentrus*, *Abudefduf*, *Chromis*, and *Dascyllus*, and it is apparent that the general behavior and spatial relationships are similar to those of the bicolor damselfish (Albrecht, 1969; Hobson, 1968; Emery, 1968; Cummings, 1968; Myrberg *et al.*, 1967; Randall, 1967; Stevenson, 1963; Eibl-Eibesfeldt, 1962; Hiatt and Strasburg, 1960; Helfrich, 1958; Albrecht and Mariscal, personal communication). Changes in activities and spatial relationships with decreased light in species of *Pomacentrus*, *Abudefduf*, *Chromis*, and *Eupomacentrus* are very similar to those observed in the bicolor damselfish (Hobson, 1968; Starck and Davis, 1966; and especially Emery, 1968). The observations cited represent areas from Florida, the West Indies, the Gulf of California, Hawaii, the Marshall Islands, and the Indian Ocean, so it is likely that the abundant pomacentrids are an important though fluctuating pathway for food and energy flow into reef ecosystems over large areas.

Many other groups of reef fishes also feed on plankton. Randall (1967), for instance, listed 20 families of reef fishes with species that feed on "zooplankton" and six families that feed "in part on animals of the plankton." He also stated that "the juveniles of many fishes such as the pomadasyids and carangids feed primarily on zooplankton," which corroborates numerous observations made by the author.

The contrast between the behavior of many species of fishes during periods of current flow as opposed to periods without current is striking. When current is absent, there is a seeming lack of orientation and organization among the fish inhabitants of the reef at Bimini. Although some species move around searching the bottom for food, a great many others idly pick at the bottom or wander around without engaging in specific behavioral activities except occasional acts of aggression or reproduction. With the onset of current, however, the complexion of the fish community becomes vastly changed. A large component of the community immediately reacts by facing into the oncoming current and by rising to varying heights above the bottom, where individual fishes begin to dart around catching planktonic organisms. This behavior makes many species readily visible, and the gradation of sizes from small to large individuals with increasing distance from the substrate can be seen. Close examination of rock and other surfaces reveals small individuals that rarely leave such confines as well as predatory species that mostly remain in contact with or close to the substrate. When evening crepuscular periods approach, most diurnal species retreat to shelter as does the bicolor damselfish, with larger individuals leaving the water column last.

The findings presented here show fundamental ways in which behavior interacts with environmental factors to regulate food intake and therefore energy inflow into a reef ecosystem. Because of their broad nature, these observations provide a basic framework to which other information pertinent to inflow can be added in further exploring the dynamics of this system. Extensive field studies of food and feeding are needed. Availability of food should be determined by extensive sampling of plankton near feeding fishes. These results could then be compared with stomach contents of fishes to determine a "field level" of selection which could be used for predicting food inflow based on a routine sampling program. More critical studies of the feeding "deflection" that have been used here as an index of feeding rate must be made with more refined viewing equipment, because fishes do not always eat food particles they take into their mouths or examine. More individual fishes should be observed so that individual variablility within particular size groups can be determined to obtain an average feeding rate.

Laboratory studies can contribute additional information that will enable us to make more critical observations of behavior seen in the field. Food selection and deprivation, in particular, must be studied quantitatively under controlled conditions, since they could show the role of hunger in regulating distance from shelter in different-sized fishes. A knowledge of these relationships might be useful for predicting the distance relationships of fishes in the field that have been deprived of food for days at a time by reduced light intensities or turbidity. Whether or not current acts as a stimulus for a conditioned response that initiates feeding could be determined as well as the energy that fishes expend in swimming against it.

Finally, observations have shown that, timewise, almost the entire day of the bicolor damselfish is spent feeding. This provides a basic framework into which other activities can be fitted in a manner that gives a more comprehensive perspective of how the animal's behavior relates to its ecology. Reproduction, for instance, is confined almost entirely to the very early morning crepuscular period. Aggression, or at least conspecific chasing, appears to occur at a much higher level during the evening crepuscular period. While feeding during periods of full daylight, larger juveniles display little aggression, although it may be higher in small individuals that remain close to shelter. The role that predation plays within this basic framework of feeding behavior has been discussed earlier.

It is hoped that this study of the bicolor damselfish will serve as a basis of comparison for much needed quantitative studies of other benthic plankton-feeding reef fishes.

ACKNOWLEDGMENTS

This research was supported by the Office of Naval Research, Oceanic Biology Program, contract: Nonr 840(13). The author wishes to thank John C. Steinberg and his staff, who developed and made operational the Bimini Underwater Video-Acoustic Installation with which this research was carried out. Special thanks are due Arthur A. Myrberg, Stanley

Walewski, John Shoup, Juanita Spires, Charles Gordon, and Theodore Crabtree, without whose advice, consultation, and help this research could not have been accomplished, and Robert F. Mathewson, Director of the Lerner Marine Laboratory, and his staff for their kind hospitality and aid. Contribution No. 1427, Rosensteil School of Marine and Atmospheric Sciences, University of Miami.

REFERENCES

Abel, E. F., 1962, Freiwasserbeobachtungen an Fischen im Golf von Neapel als Beitrag zur Kenntnis ihrer Ökolgie und ihres Verhaltens, *Intermat. Rev. Ges. Hydrobiol.* **47(2)**: 219–290.

Albrecht, H., 1969, Behavior of four species of Atlantic damselfishes from Colombia, South America, *Abudefduf saxatiles*, *A. taurus*, *Chromis multilineata*, *C. cyanea* (Pisces, Pomacentridae), *Z. Tierpsychol.* **26**: 662–676.

Bardach, J. E., 1958, On the movements of certain Bermuda reef fishes, *Ecology* **39(1)**: 139–146.

Breder, C. M., Jr., 1946, An analysis of the deceptive resemblances of fishes to plant parts, with critical remarks on protective coloration, mimicry and adaptation, *Bull. Bingham Oceanogr. Collect.* **10(2)**: 49 pp.

Cummings, W. C., 1968, Reproductive habits of the sergeant major, *Abudefduf saxatilis* (Pisces, Pomacentridae) with comparative notes on four other damselfishes in the Bahama Islands, Doctoral dissertation, University of Miami, 172 pp.

Eibl-Eibesfeldt, I., 1962, Freiwasserbeobachtungen zur Deutung des Schwarmverhaltens verschiedener Fische, *Z. Tierpsychol.* **19(2)**: 165–182.

Emery, A. R., 1968, Comparative ecology of damselfishes (Pisces: Pomacentridae) at Alligator Reef, Florida Keys, Doctoral dissertation, University of Miami, 258 pp.

Gosline, W. A., and Brock, V. E., 1960, "Handbook of Hawaiian Fishes," University of Hawaii Press, Honolulu, 372 pp.

Helfrich, P., 1958, The early life history and reproductive behavior of the maomao, *Abudefduf adbominalis* (Quoy and Gaimard), Doctoral dissertation, University of Hawaii, 228 pp.

Hiatt, R. W., and Strasburg, D. W., 1960, Ecological relationships of the fish fauna on coral reefs of the Marshall Islands, *Ecol. Monogr.* **30**: 65–127.

Hobson, E. S., 1965, Diurnal–nocturnal activity of some inshore fishes in the Gulf of California, *Copeia* **(3)**: 291–302.

Hobson, E. S., 1968, Predatory behavior of some shore fishes in the Gulf of California, U.S. Fish Wildlife Serv., Res. Rep. No. 73, 92 pp.

Longley, W. H., Schmitt, W. L., and Taylor, W. R., 1925, Observations upon the food of certain Tortugas fishes, *Ann. Rep. Tortugas Lab. Carnegie Inst. Wash.* **24**: 230–232.

Myrberg, A. A., Jr., Brahy, B. D., and Emery, A. R., 1967, Field observations on reproduction of the damselfish, *Chromis multilineata* (Pomacentridae), with additional notes on general behavior, *Copeia* **(4)**: 819–827.

Randall, J. E., 1961, Tagging reef fishes in the Virgin Islands, Proceedings of the Caribbean Fisheries Institute, 14th Annual Session, pp. 201–241.

Randall, J. E., 1967, Food habits of reef fishes of the West Indies, *Stud. Trop. Oceanogr., Inst. Mar. Sci. Univ. Miami* **(5)**: 665–847.

Starck, W. A., and Davis, W. P., 1966, Night habits of fishes of Alligator reef, Florida, *Ichthyol. Aquarium J.* **38(4)**: 313–356.

Stevenson, R. A., 1963, Life history and behavior of *Dascyllus albisella* Gill, a pomacentrid reef fish, Doctoral dissertation, University of Hawaii, 221 pp.
Suyehiro, Y., 1942, A study on the digestive system and feeding habits of fish, *Japan. J. Zool.* **10(1)**: 1–301.
Wickler, W., 1967, Specializations of organs having a signal function in some marine fish. *Stud. Trop. Oceanogr., Inst. Mar. Sci. Univ. Miami* **(5)**: 539–548.
Wiens, H. J., 1962, "Atoll Environment and Ecology," Yale University Press, New Haven, Conn, 532 pp.
Woodhead, P. M. J., 1966, The behavior of fish in relation to light in the sea, *Oceanogr. Mar. Biol. An. Rev.* **4**: 337–403.

Chapter 8

DAILY AND SEASONAL RHYTHMS OF ACTIVITY IN THE BLUEFISH (*POMATOMUS SALTATRIX*)

Bori L. Olla and Anne L. Studholme

U. S. Department of Commerce
National Oceanic and Atmospheric Administration
National Marine Fisheries Service
North Atlantic Coastal Fisheries Research Center
Laboratory for Environmental Relations of Fishes
Highlands, New Jersey

I. INTRODUCTION

The behavior of an animal, as well as its related physiological processes, is often rhythmic. An event may take place only at a certain time of day or night or season or year corresponding to certain natural external cycles. These cycles may exert varying degrees of control; their influence may be total, with the entire response dependent on the continued presence of the stimulus, or they may act as a cue, initiating, terminating, or synchronizing a particular event which is basically internally controlled. Knowledge of the relation of the external cycle to the behavior of the animal is essential to understanding the harmony between organism and environment. It allows us to predict when and where certain events are most likely to occur as well as when an animal may be predisposed to respond to a particular stimulus. In this chapter we examine, quantitatively, daily and seasonal rhythms in swimming speed and schooling of a marine pelagic migrating species, the bluefish, *Pomatomus saltatrix* L., held under controlled laboratory conditions.

There have been numerous studies of activity rhythms in a variety of organisms; many of these are highly quantitative, examining in fine detail the external and internal components of rhythmicity (see Cold Spring Harbor Symposium on Biological Clocks, 1960; Cloudsley-Thompson, 1961; Harker, 1964; Bünning, 1964; Aschoff, 1965). However, studies on marine fishes have primarily been qualitative, emphasizing the initiation and cessation of activity without quantitatively examining changes in response. As Woodhead (1966) points out, most determine only the duration of activity and not the degree of change.

In one of the few studies in which daily rhythmicity was quantitatively examined, Kruuk (1963) measured, under controlled laboratory conditions, the activity cycle of the common sole, *Solea vulgaris*. Under a normal day–night cycle, the fish showed a daily rhythm of swimming activity, with maximum movement at night. When the fish were subjected to constant dark, the rhythm persisted for about two cycles, suggesting some degree of internal control, although under constant light this periodicity seemed to disappear. Under an imposed 6-hr light–dark regimen, the rhythm corresponded to the imposed light cycle; however, the fish were most active during the period when the imposed darkness came during the natural night. Kruuk suggested that while light was important in regulating activity, another control mechanism was probably operating.

In a laboratory study on another flatfish, the plaice, *Pleuronectes platessa*, de Groot (1964) subjected the fish to artificial 6-hr light–dark periods. His preliminary findings showed some evidence that rhythmic activity was controlled internally.

Winn *et al.* (1964) working on sun-compass orientation in parrot fishes, *Scarus guacamaia* and *S. coelestinus*, shifted the activity rhythm of fishes in the laboratory by delaying the light cycle by 6 hr. When the fishes were released at sea, there was a deviation in the direction of expected movement relative to the 6-hr shift. This indicated internal control of rhythmicity and pointed to its use as an internal chronometer.

An analysis of the seasonal activity changes of migratory fishes should take into account the probable interrelationship of external and internal factors. While the exact role that each may play in both migratory and reproductive behavior is not altogether clear, Liley (1969) points out that "it seems reasonable to suppose that hormone-induced changes in responsiveness to general or specific external stimuli are in part responsible for increased locomotory activity. Such changes in activity and responsiveness may be important components in migratory and other reproductive behavior in fishes."

Studies which examined the relation between external and internal factors were performed by Baggerman (1957, 1959). Using changes in salinity preference as an indication of migration disposition of the three-spined stickleback, *Gasterosteus aculeatus* L., she observed that gonadal changes occurred simultaneously with changes in salinity preference. However, she could still alter salinity preference in gonadectomized fish by varying photoperiod and temperature, indicating that gonadal maturation was not essential to the onset of migration disposition; the external cues, in this case, apparently stimulated the change. In later work Baggerman (1962) concluded that increasing photoperiod may act by stimulating thyroid activity, which in turn affects both salinity preference and locomotor activity associated with migration.

Both reproductive and migratory success depend on precise timing. Essential to this timing is some external cue which changes seasonally, in the same way, from year to year. Two obvious external factors which might act as cues for fish residing in the photic zone are photoperiod and temperature. As an example of the former, Woodhead and Woodhead (1965), studying the Barents Sea cod, *Gadus morhua*, concluded that since no significant temperature change occured there in September when migration began, it must have been the changing light regimen which stimulated internal physiological changes resulting in migration. The first step in examining the relation of changing photoperiod to migration would be to measure daily activity under a steadily changing natural seasonal light regimen under controlled environmental conditions. This would effectively reduce the influence of other factors such as changing temperature and salinity as possible stimuli for initiating seasonal changes in activity. However, this approach would (1) not preclude the possibility of other stimuli affecting the seasonal cycle or (2) negate the presence of a cycle wholly internally controlled.

There is an indication that both activity rhythms and retinal adaption, while correlated with the light cycle to a certain extent, may be under some degree of internal control. Several freshwater and anadromous fish species kept in constant darkness exhibited rhythms of photomechanical changes in the retina, implying an internal control mechanism (Welsh and Osborn, 1937; Wigger, 1941; Arey and Mundt, 1941; Ali, 1961; John and Haut, 1964; John *et al.*, 1967; John and Gring, 1968; John and Kaminester, 1969). Olla and Marchioni (1968) reported the results of a study on rhythmic cone movement in retinas of bluefish held under constant darkness. They found that swimming activity and photomechanical changes in the retina were related to light, were diurnal, and were under some degree of internal control. Those findings will be summarized in the *Results* section of this chapter.

Our objectives in this chapter are to examine in the bluefish (1) whether there is a daily rhythm of activity as evidenced by changes in swimming speed and schooling-group size during the day and night and if so what are the rhythmic characteristics, (2) whether there is a relation between daily activity and changing seasonal photoperiod, (3) what is the role of external and internal factors in rhythmic control, and (4) what is the relation of adaptive changes in the retina to external and internal factors and to the daily rhythm of activity.

II. MATERIALS AND METHODS

The subject of our observations, six adult bluefish, ranged in length from 45 to 55 cm. We held the fish under controlled light and temperature in a multiwindowed elliptical seawater aquarium 10.6 by 4.5 by 3 m with a capacity of 121 kiloliters (Olla *et al.*, 1967). A sand and gravel filtration system assured good water quality and clarity; pH ranged from 7.1 to 7.7 and oxygen from 6.8 to 7.5 ppm; salinity averaged about 24‰. Water temperature ranged from 18.0 to 20.0°C.

A specialized lighting system simulated diurnal changes in the light intensity and duplicated natural seasonal photoperiod. Eight rows of fluorescent lights (color temperature 5500°K), mounted on each side wall above the aquarium and controlled by a series of timers, simulated day length from morning civil twilight to evening civil twilight (Table I). An automated dimming system controlled the top row, causing light to come on gradually to avoid sudden changes that would startle the fish. In the evening the process reversed, with the dimmers effecting gradual onset of the dark period. A series of incandescent bulbs contained in boxes covered with diffusing screens produced the simulated night-time light at a level of 0.02 ft-candle. These boxes pointed upward, causing light to reflect indirectly off the walls

Table I. Cumulative Light Intensity Values of the Lighting System Measured at Water Surface

Light period	Light source		Total light intensity (ft-candles)	
Dark	Incandescent lights (light boxes)		0.02	
Morning and evening simulation of civil twilight	Top row of fluorescent lights		0.02 Morning 37[a] ⇄ Evening	
	1–2 rows		62	
	1–3 rows		100	
	1–4 rows		150	
Light	1–5 rows	Morning ↓	200	Evening ↑
	1–6 rows		250	
	1–7 rows		300	
	1–8 rows		350	

[a] Times of light and dark onset are automatically adjusted to coincide with the natural seasonal photoperiod.

and ceiling. A time switch turned on these bulbs before the dimmer lights became extinguished. Daily photoperiod coincided with the current natural photoperiod except when the light regimen was experimentally shifted. During experiments under constant light, we set the top row at the prescribed intensity level, and during those under shifted light, we advanced the total light period the required number of hours from normal.

We used swimming speed as one indicator of activity throughout the study. Typically, the bluefish would swim in one or more groups around the tank, 30–60 cm from the sides, allowing us to clock the time for them to pass between two marks 335 cm apart. Every hour beginning on the hour (except for one day when we took two readings every 5 min), we made five stop watch readings of the time taken for the leading fish of the largest group to swim this distance. We converted these measurements to speed (cm/sec) and for subsequent analysis used the median of the five readings, thus eliminating irregular speeds which occurred from time to time within the five readings.

As another indicator of activity, we used schooling-group size. We used the number of fish grouped as a measure of schooling and considered the fish to be grouping (schooling) when two or more individuals were similarly oriented, swimming in the same direction at relatively the same speed within three body lengths. Where there was no grouping, i.e., the fish were swimming independently of one another, the schooling-group size was designated 0. We recorded five counts of the largest number of fish grouped at the same time as we measured speed and used the median for subsequent analysis.

We observed activity under varying photoperiods in 4-day sets. Since preliminary observations showed that the presence of live prey in the tank disrupted the pattern of daily activity, we fed the fish to satiation following each 4-day set of observations and allowed a 3-day recovery period from the feeding before beginning the next set of observations. We fed the fish to satiation 3 days before experiments in which light was shifted or held constant and did not feed them again until the completion of the series.

III. RESULTS

A. Daily Activity

Examination of the accumulated data on the swimming speed and schooling activity of the bluefish showed certain characteristic patterns in their daily activity. A representative 4-day period (photoperiod 14.15 hr) with readings recorded hourly and 1 single day (photoperiod 12.50 hr) with

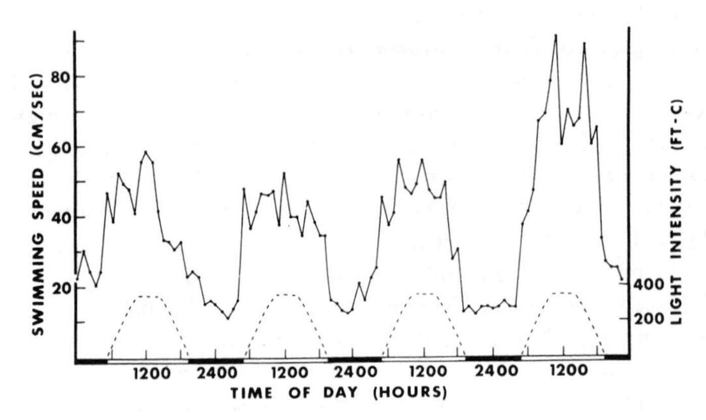

Fig. 1. Hourly median swimming speeds measured for 4 days under a 14.15-hr photoperiod. Changing light intensity is indicated by broken lines.

readings recorded at 5-min intervals illustrate the basic features of daily activity.

1. Swimming Speed

Most obvious from examining the hourly median measurements of swimming speed for 4 consecutive days was that speed varied rhythmically, with a period of about 24 hr (Fig. 1).

A primary characteristic of the rhythm was a sharp increase in speed from the last hourly reading before light onset to the first reading after light onset which was greater than any other increase throughout the day. In this particular 4-day example, the first speed measurements during the light period followed light onset by 6 min.

In the ensuing 6–7 hr after light onset, the speed gradually but consistently increased. During the time corresponding to afternoon, speed decreased rapidly. This decrease tapered off within 1–2 hr after dark onset. During the dark phase, the fish maintained their decreased swimming speed until morning light onset.

The swimming rate varied more during the dark phase than during the light phase. The coefficient of variation (c.v.*) was shown to be 28.2% for the dark period and only 12.7% for the light period (Table II).

When we measured swimming speed every 5 min for 24 hr (Fig. 2), we found a high degree of variation between readings, superimposed on a basic 24-hr rhythm. An upward trend in swimming speed began 10 min

* c.v. $= \sigma'/\bar{x} = \bar{R}_4/1.978\,\bar{x}$, where $\bar{x}$ is a grand average speed. See Ferrell (1958) for derivation of σ'.

Table II. Short-Term Variation in Swimming Time as Measured by the Hourly Range (in sec) of Five Replicate Stop Watch Readings Measured for 4 Days Under a 14.15-hr Photoperiod

	Day			
Hour	1	2	3	4
		Dark		
0	13.2	22.1	9.2	11.6
1	21.9	16.1	18.3	7.4
2	16.9	10.6	22.9	9.6
3	14.2	10.3	3.8	10.4
4	18.6	9.4	4.0	10.6
		Light		
5	1.3	2.8	1.3	1.8
6	1.7	2.9	4.4	1.2
7	0.8	1.8	2.6	0.8
8	0.6	1.8	1.3	0.8
9	2.0	2.6	3.4	1.1
10	3.4	1.5	2.9	1.3
11	2.2	2.1	2.7	1.0
12	1.2	1.1	0.6	0.6
13	1.6	1.8	4.9	1.0
14	4.6	2.0	2.3	2.4
15	5.6	1.6	3.3	1.1
16	2.2	1.2	2.0	1.2
17	2.8	2.4	9.0	1.3
18	3.7	3.2	7.8	1.6
19	12.7	2.0	20.0	4.8
		Dark		
20	4.0	9.6	7.5	17.2
21	27.2	7.6	27.4	8.0
22	15.2	12.6	5.2	5.4
23	12.3	13.8	28.0	18.9

	Light	Dark
$\tilde{R}$ (median range in sec)	2.0	12.0
$\bar{x}$ (mean swimming time in sec)	7.1	18.8
σ' (estimated standard deviation of the residual dispersion of stop watch readings $= \tilde{R}/2.257$)	0.9	5.3
c.v. ($\sigma'/\bar{x}$)	12.7%	28.2%

after light onset, while the greatest change from the low speed observed during the dark phase occurred 20–25 min after light onset.

2. Schooling

The measurements made for 4 days on hourly schooling-group size reflected a distinct day–night change in the schooling response of the fish corresponding to the swimming–speed rhythm (Fig. 3). During the day there was a high degree of group interaction, the fish acting in most instances as

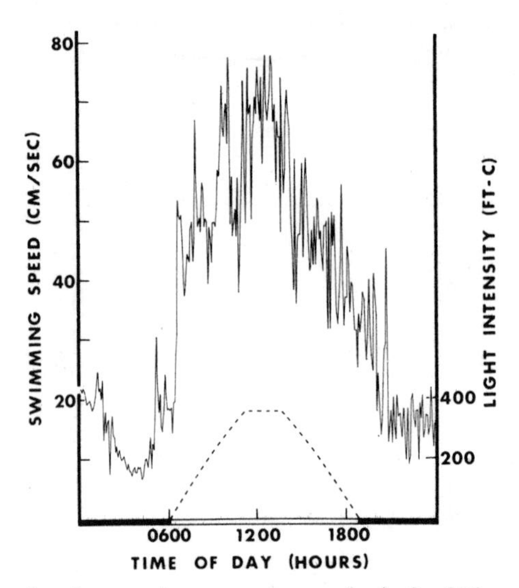

Fig. 2. Median swimming speeds measured every 5 min for 24 hr under a 12.50-hr photoperiod. Changing light intensity is indicated by broken lines.

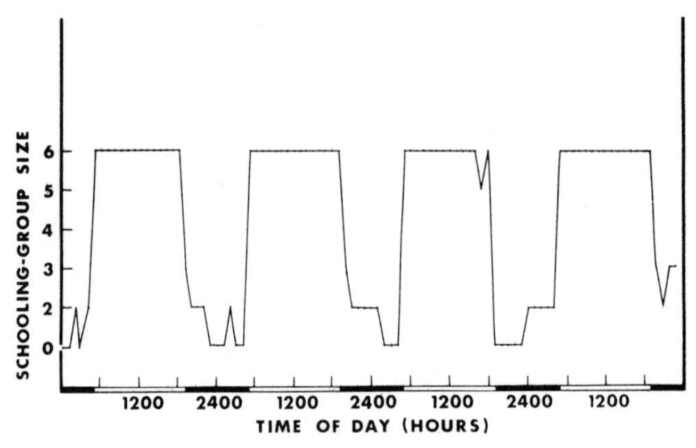

Fig. 3. Hourly medians of schooling-group size measured for 4 days under a 14.15-hr photoperiod.

a single schooling unit. At night the tendency for grouping was significantly lower.

A change from varying schooling-group sizes at night to a consistently high group size during the day was evident from schooling measurements taken every 5 min, 2 hr before to 2 hr after morning light onset (Fig. 4). This change occurred 5–10 min after light onset.

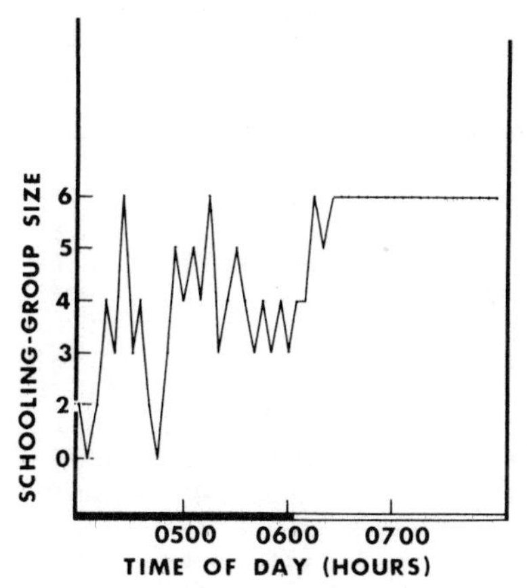

Fig. 4. Medians of schooling-group size measured every 5 min from 2 hr before to 2 hr after light onset.

B. Seasonal Activity

We observed the fish under normal day–night cycles and constant temperature during different seasons to examine whether there were any consistent changes in daily activity corresponding to changes in the natural seasonal photoperiod. We measured hourly changes in swimming speed and schooling in the same manner as previously described at 19 different nonsequential photoperiods from 10.30 hr to 16.16 hr for a total of 76 days measured over an 18-month period. Analysis of our data chronologically showed that growth in body length during this time had a negligible effect on the results.

1. Swimming Speed

We studied swimming speed in 4-day sets at 19 different photoperiods (Table III). A 4-day set (each column) consisted of 24 medians of four hourly median speeds (i.e., the median of the four median speeds at 0100, 0200, etc.).

The basic diurnal rhythm of swimming speed was apparent at all the photoperiods examined. This was confirmed by the Tukey–Duckworth end-count test (Tukey, 1959) for each of the 19 sets, in which even the smallest total end-count indicated a high level of significance ($P = 0.001$).

To study the relationship between photoperiod and average speed, we divided the hourly median speeds of each column in Table III into two 7-hr groups, day (0900–1500) and night (2100–0300), and computed their means. We limited the analysis to 7-hr periods to avoid the problem of over-

Table III. Hourly Median Speeds (in cm/sec)

Hour		Photoperiod 10.30	10.66	10.73	10.80	11.50	11.60	11.87	12.03	12.10
0		10.2–[a]	9.6–	26.3	8.9–	29.3–	14.6–	15.7–	18.6–	12.6–
1		11.3–	6.6–	20.7–	12.2–	23.5–	12.8–	13.7–	16.0–	12.4–
2		9.1–	8.0–	18.9–	11.4–	23.7–	13.6–	11.7–	22.4–	14.7–
3	Dark	10.9–	8.2–	19.4–	13.0–	24.1–	14.6–	17.6–	21.0–	14.0–
4		9.6–	8.1–	22.6–	15.6–	31.0	15.0–	15.1–	17.2–	11.6–
5		13.8–	7.5–	19.4–	18.8–	25.0–	14.8–	20.2–	21.2–	15.0–
6		11.3–	9.7–	23.0–	17.5–	23.7–	19.0–	26.0	13.9–	22.2
7		26.4+	29.6+	38.5+	42.0+	36.8+	36.0+	39.5+	32.7+	36.5+
8		28.0+	31.4+	42.4+	45.6+	42.5+	39.7+	38.3+	36.0+	39.8+
9		34.9+	34.7+	56.8+	38.0+	44.1+	34.9+	41.9+	41.1+	38.4+
10		35.8+	33.3+	47.9+	42.5+	48.9+	35.8+	45.3+	36.4+	34.1+
11		39.8+	30.6+	42.0+	37.5+	46.4+	36.8+	49.7+	40.4+	35.6+
12	Light	36.8+	28.6+	45.6+	45.3+	44.4+	33.8+	47.0+	31.7+	31.3+
13		45.1+	28.3+	43.3+	44.1+	42.4+	35.1+	45.9+	30.7+	30.2+
14		39.0+	23.0+	33.4+	37.2+	37.0+	34.4+	38.1+	32.2+	20.9
15		30.6+	22.4+	32.0+	35.6+	34.1+	28.7+	47.2+	27.5+	26.6+
16		29.1+	24.1+	23.0	22.5+	30.7	29.6+	37.8+	27.6+	26.3+
17		19.1+	27.9+	25.9	20.5–	36.1+	29.6+	34.3+	23.6+	25.5+
18		13.2–	14.5–	27.8	19.7–	34.7+	21.4–	21.9	32.1+	21.3
19		11.9–	11.7–	29.4	18.2–	21.8–	18.1–	23.5	17.6–	13.8–
20	Dark	10.3–	12.2–	25.7	14.6–	22.0–	19.0–	16.2–	20.1–	14.6–
21		7.8–	11.5–	23.2	17.1–	28.1–	18.1–	16.6–	13.7–	12.6–
22		9.2–	11.4–	27.3	14.0–	25.7–	17.4–	13.4–	12.8–	12.0–
23		11.2–	8.1–	26.7	10.9–	27.9–	16.0–	16.1–	17.5–	16.5–
$\bar{x}_{24}$		21.0	18.4	30.9	25.1	32.7	24.5	28.9	25.2	22.4
End-count		24	24	15	24	22	24	21	24	21
P		10^{-6}	10^{-6}	10^{-3}	10^{-6}	10^{-5}	10^{-6}	10^{-5}	10^{-6}	10^{-5}

[a] + = Day values equal to or larger than the largest night value. – = Night values equal to or

lapping day–night speeds due to different light and dark onset times. We subsequently converted the means to $\log_e$ in order to stabilize the variance about the mean swimming speeds and plotted these against photoperiod (Fig. 5); regression lines were then fitted to day and night groups.

Analysis of variance for linear regression indicated a positive correlation between speed and changing photoperiod for both day ($F = 425.3$; 1, 114 df) and night ($F = 154.3$; 1, 114 df). In addition, the lines describing day and night both have a common slope ($F < 1$); this would indicate that on a percentage basis changing photoperiod equally affected swimming speed both day and night.

Rhythm amplitude, like average speed, was greater at the longer photoperiods. Applying Fourier analysis to each of the 19 sets and plotting the amplitudes (Fig. 6), we found a significant difference between amplitudes measured at the long (13.75–16.16 hr) and short (10.30–12.10 hr) photoperiods ($P = 0.01$, Tukey–Duckworth end-count test).

Measured During 4-Day Tests under 19 Different Photoperiods

(hr)									
13.75	14.00	14.18	14.70	15.07	15.33	15.37	15.50	15.90	16.17
20.5−	23.1−	13.9−	19.7−	19.3−	31.9−	25.6−	21.3−	25.1−	29.4−
23.1−	26.4−	17.1−	17.4−	16.3−	11.5−	23.8−	20.4−	21.7−	23.8−
21.6−	21.6−	15.8−	26.5−	21.5−	11.3−	26.7−	23.7−	21.3−	19.2−
20.0−	21.1−	17.1−	25.1−	18.3−	10.4−	31.0	22.0−	18.7−	24.7−
21.9−	24.3−	20.1−	23.1−	24.3−	15.5−	26.8−	26.0−	22.1−	15.7−
30.4−	37.9+	45.9+	44.1+	48.5+	32.2	43.2+	50.9+	39.9+	46.1+
52.0+	58.6+	37.8+	47.9+	53.6+	39.2+	59.9+	53.4+	42.6+	50.9+
45.9+	57.4+	43.9+	52.3+	59.6+	44.4+	52.1+	57.0+	45.6+	54.9+
53.9+	57.3+	52.5+	52.3+	53.5+	48.5+	48.4+	53.8+	55.0+	52.8+
56.3+	58.9+	47.9+	48.2+	56.8+	48.4+	50.0+	49.8+	58.0+	56.5+
52.8+	56.6+	46.5+	47.0+	71.5+	56.3+	52.9+	68.4+	56.0+	61.1+
50.8+	55.5+	52.2+	48.8+	61.6+	51.3+	54.0+	62.3+	65.7+	54.9+
52.2+	62.4+	52.3+	55.0+	62.7+	50.4+	52.4+	67.2+	73.8+	59.4+
60.5+	63.8+	51.5+	55.0+	87.8+	50.8+	61.4+	80.7+	73.6+	61.5+
54.4+	49.3+	43.3+	50.8+	72.7+	60.8+	53.3+	69.6+	80.7+	69.2+
52.8+	59.2+	39.4+	53.5+	80.0+	58.6+	50.6+	83.7+	70.6+	70.4+
53.6+	51.6+	46.7+	48.2+	71.1+	50.1+	48.6+	58.4+	72.5+	54.9+
40.4+	44.4+	34.4+	42.4+	50.3+	52.5+	43.4+	44.7+	49.6+	38.2+
34.5+	45.4+	33.0+	40.2+	40.7+	43.8+	39.0+	34.9+	39.0+	35.3+
24.5−	32.7−	27.7+	38.5+	34.4+	38.1+	29.9	31.5+	38.0+	30.3+
22.9−	24.9−	20.0−	22.8−	23.5−	34.5	21.5−	29.2−	27.4	31.7+
25.2−	23.0−	18.9−	15.3−	21.2−	26.8−	24.6−	27.1−	36.1	22.8−
19.5−	24.6−	14.4−	13.8−	20.2−	28.3−	20.8−	26.7−	35.9	25.0−
19.2−	24.7−	15.1−	19.2−	19.8−	26.1−	23.1−	26.9−	36.2	20.2−
37.9	41.9	33.6	37.8	45.4	38.4	40.1	45.4	46.0	42.0
24	24	24	24	24	22	22	24	20	24
10^{-6}	10^{-6}	10^{-5}	10^{-5}	10^{-5}	10^{-5}	10^{-5}	10^{-5}	10^{-4}	10^{-5}

smaller than the smallest day value.

To examine the day-to-day variability in speed as related to length of photoperiod, we calculated the coefficient of variation for light and dark periods for short (10.30–12.10 hr) and long (13.75–16.16 hr) photoperiods. For daytime readings at the long photoperiods this proved to be 21.5%; for night-time readings 32.8%. For daytime readings at the short photoperiods it was 21.3%; for night-time readings 32.5%. These percentages signified that the swimming speed varied more in the dark than in the light at all photoperiods and that there was no difference in this variability throughout the year.

2. Schooling

We examined the differences in schooling tendency between day and night at each of the 19 photoperiods by recording the number of fish in the largest group five times each hour. We summarized the data by calculating the occurrence of each group size (0, 2–6) and expressed these as percentages of the total dark or light counts (Table IV). During the light period, the highest percent always occurred in group-size 6; i.e., the fish showed the strongest tendency to school. During the dark, with four exceptions, the highest percent occurred in group-size 0; i.e., the fish were swimming independently of one another. Figure 7 shows the mean percent of both dark and light observations for all 19 photoperiods, clearly indicating that the tendency to group was greater during the day than at night.

Examining the effect of photoperiod on schooling-group size, we used the group-size 0, "dark" column, and the group-size 6, "light" column, as indicative of the behavior of the fish most of the time. Using the runs test (Olmstead, 1958), we found no change in grouping tendency during the dark and a slight increase during the light period under the long photoperiod ($P = 0.05$).

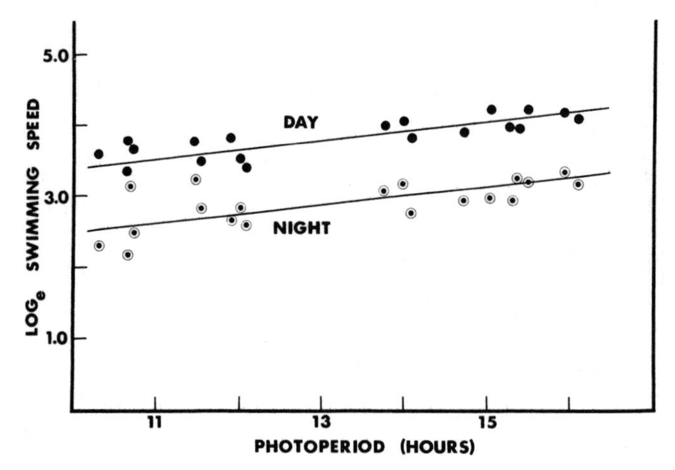

Fig. 5. Mean day and night swimming speeds of 4-day sets measured at 19 different photoperiods.

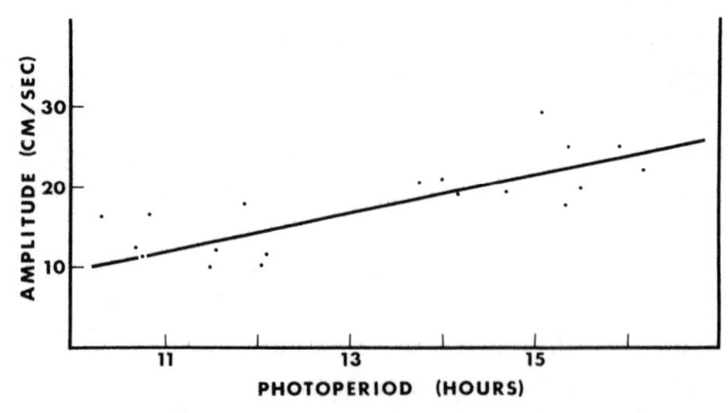

Fig. 6. Rhythm amplitudes of 4-day sets measured at 19 different photoperiods.

Table IV. Percent Occurrence of Each Schooling-Group Size under Light and under Dark for 19 Different Photoperiods

	Schooling-group size											
	Dark						Light					
Photoperiod (hr)	0[a]	2	3	4	5	6	0	2	3	4	5	6
					Short							
10.30	56.4[b]	21.9	8.5	6.2	3.2	3.8	4.0	6.2	4.9	6.2	13.5	65.2
10.66	78.1	8.5	7.4	2.6	0.8	2.6	12.2	0.5	0.5	0.9	1.4	84.5
10.73	23.5	26.1	12.3	13.5	8.8	15.8	6.8	5.9	4.1	7.7	12.3	63.2
10.80	48.9	22.1	15.0	7.5	5.4	1.1	12.0	8.0	11.5	10.0	12.5	46.0
11.50	6.3	4.6	5.4	2.9	5.0	75.8	0	0	0.1	1.7	1.2	97.0
11.60	44.0	27.7	14.2	8.8	3.8	1.5	0.9	0.9	3.0	2.0	10.5	82.7
11.87	39.2	31.2	15.0	7.9	4.6	2.1	1.7	1.7	2.5	4.2	9.1	80.8
12.03	25.8	12.1	11.3	4.2	10.8	35.8	0.8	0.5	0.8	0	0.8	97.1
12.10	48.3	22.1	13.3	6.7	7.1	2.5	9.6	6.2	6.7	9.2	10.0	58.3
					Long							
13.75	29.1	19.6	11.4	10.9	7.7	21.3	0	0	0.4	0.9	3.4	95.3
14.00	37.5	18.5	13.0	10.5	5.5	15.0	1.0	1.7	0.8	0.8	2.5	93.2
14.18	46.0	26.0	18.0	8.3	0.6	1.1	2.3	0	1.0	0.7	4.0	92.0
14.70	50.0	26.7	12.2	6.1	3.9	1.1	0.3	0	0.3	1.7	6.2	91.5
15.07	61.1	23.8	7.2	3.3	2.3	2.3	0	0.3	0.3	1.1	3.3	95.0
15.33	10.0	6.7	17.8	17.8	20.5	27.2	0	0.3	1.4	0.3	6.0	92.0
15.37	58.3	23.9	6.7	5.6	2.2	3.3	0.3	1.7	0.3	3.0	4.7	90.0
15.50	60.7	22.2	10.0	3.8	2.2	1.1	5.0	3.0	2.0	3.0	4.0	83.0
15.90	38.1	18.8	15.0	11.3	1.8	15.0	5.0	3.4	2.5	2.5	3.1	83.5
16.17	33.8	28.1	14.4	4.4	5.6	13.7	0.9	0.9	2.8	2.8	3.8	88.8
Mean percent	41.8	20.6	12.0	7.5	5.4	12.7	3.3	2.2	2.4	3.1	5.9	83.1

[a] 0 indicates measurements were made on one fish.
[b] Underline indicates highest percent in each photoperiod under dark conditions and under light conditions.

C. Shifted Light

1. Swimming Speed

From the observations described, it was clear that the daily rhythm of activity correlated closely with the daily cycle of light. However, up to this point we could not specifically state what role the light played in daily rhythmicity. The rhythm could be a response to daily fluctuations in light or a response to some other undefined external stimulus which was correlated with the light cycle. On the other hand, the rhythm could be internally controlled, with some external cue, light in this case, acting as a phasic component.

To examine these possibilities we first subjected the fish to a shifted cycle of light in which the times of light and dark onset were delayed by 7 hr

(Fig. 8). If the swimming-speed rhythm followed the shift immediately, i.e., with no significant increase at real sun time but only at the onset of the delayed light cue, we could conclude that the rhythm was externally controlled by light acting as the primary stimulus. However, if there were a lag in the rhythm conforming to the light shift, i.e., if swimming rate accelerated at real sun time even without a light cue and if the activity pattern gradually conformed to the shifted light cycle, we would assume that there was internal control of rhythmicity with light acting as a phasic component.

Table V shows the hourly median speeds measured 1 day before, during, and 6 days after a 7-hr delay in the light cycle. To show in detail the results of experiments with shifted and constant light, we have presented a nonparametric analysis of our data in tabular form. Using the runs test (Olmstead, 1958) to indicate significant daily positive and negative runs (values above and below the median), we found that on the day of the light delay (day 2), although light intensity remained at 0.02 ft-candle at the time of normal light onset, the fish swam at high day speeds. They also showed no appreciable response to the delayed light onset. Swimming speed corresponded closely to the preceding normal light cycle. On day 3, the second day of the light shift, there was variability in swimming speed, with a response to light onset, but no sustained rhythmic trend. By the third day of the shift (day 4) and for the next 4 days, the fish showed a well-

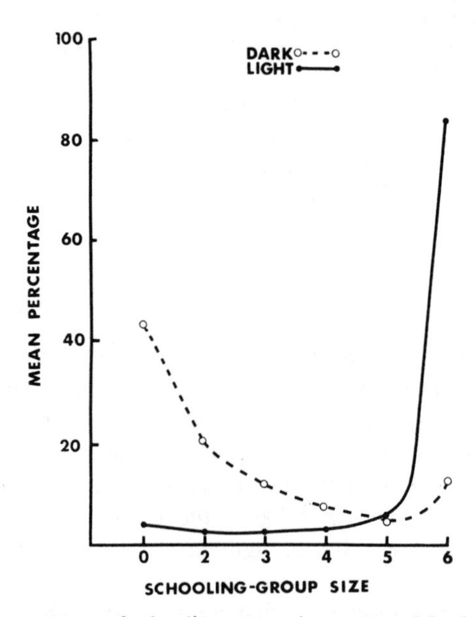

Fig. 7. Mean percentage of schooling-group size measured for both dark and light periods under 19 different photoperiods.

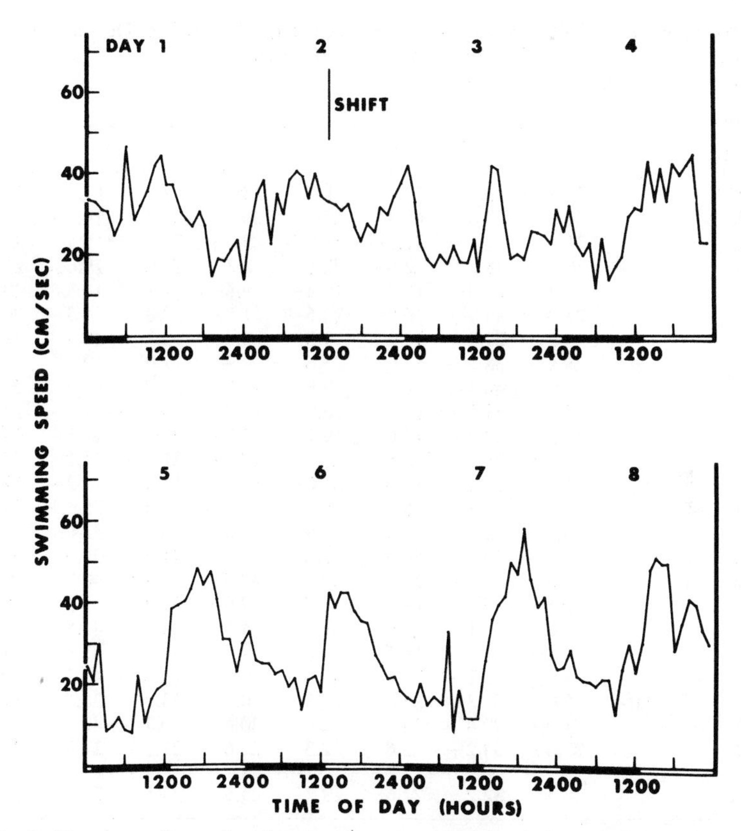

Fig. 8. Hourly median swimming speeds measured 1 day before, during, and 6 days after a 7-hr light delay.

defined rhythm in swimming speed corresponding to the shifted light cycle. The lag in conforming to the shifted cycle, followed by in-phase conformance, led us to conclude that (1) there was an internal component in the rhythm and (2) light played a major role in phasing the rhythm.

To examine further the phasic role of light and the lability of the rhythm, we delayed the light cycle in three steps of 7 hr each. Between each light shift we waited several days until the rhythm conformed to each shifted cycle.

Table VI shows the hourly median speeds beginning 6 days after the initial 7-hr light delay, by which time the daily activity rhythm corresponded to the shifted light period. Immediately following the second and third shifts, 14 and 21 hr, respectively, from the normal light cycle, the fish required 1–2 days to adapt. During these periods of adaptation, vestiges of the previous rhythm were evident, as shown by the increase in speed on days 4 and 8 before light onset on days 4 and 9. These results further con-

Table V. Hourly Median Speeds (cm/sec) Measured 1 Day Before, During, and 6 Days after a 7-hr Light Delay[a]

Hour		Day							
		1	2	3	4	5	6	7	8
0		33.5+[b]	13.8−	37.2+	31.0+	24.3+	29.4+	17.6−	23.3−
1		32.8+	27.0−	41.9+	25.7−	20.7−	32.2+	16.3−	23.9−
2		31.0+	34.9+	32.8+	31.6+	29.6+	25.4+	15.4−	27.9−
3	Dark	30.5±	38.1+	22.6±	22.6−	8.3−	24.6+	19.6−	21.8−
4		24.8−	22.5−	18.8−	19.8−	9.6−	24.3−	14.3−	20.7−
5		28.4−	34.9+	16.9−	22.6−	11.8−	22.0−	16.5−	20.2−
6		46.5+	29.9−	19.9−	11.3−	8.6−	22.9−	14.8−	19.3−
7		28.4−	38.1+	17.6−	23.6−	7.9−	19.0−	32.2+	21.0−
8		31.9+	40.4+	22.0−	13.8−	21.9−	20.7−	7.6−	21.0−
9		35.6+	39.0+	17.8−	16.9−	10.3−	13.2−	18.0−	12.7−
10		41.9+	33.5+	17.6−	19.1−	16.0−	20.6−	11.1−	23.6−
11		44.1+	39.9+	23.6+	28.9−	18.4−	21.3−	11.1−	29.4+
12	Light	37.2+	34.2+	15.6−	31.0+	19.7−	17.5−	11.2−	22.9−
13		37.2+	32.8+	28.4+	30.5+	38.1+	41.9+	25.4+	29.9+
14		30.5±	32.2±	41.9+	42.9+	39.0+	38.1+	35.6+	47.9+
15		28.4−	30.5−	40.9+	32.8+	39.9+	41.9+	39.0+	50.8+
16		27.0−	32.2±	27.9+	40.9+	42.9+	41.9+	40.9+	49.3+
17		30.5±	26.6−	18.8−	32.8+	47.9+	37.2+	49.3+	49.3+
18		27.0−	23.1−	19.9−	41.9+	44.1+	34.9+	46.5+	28.1+
19		14.6−	27.5−	18.6−	39.0+	47.2+	34.2+	57.8+	34.9+
20		18.8−	25.2−	25.4+	41.9+	40.4+	26.6+	45.3+	40.9+
21	Dark	18.0−	31.3−	25.0+	44.1+	30.5+	23.9−	38.5+	39.4+
22		21.0−	29.4−	24.3+	22.6−	30.5+	20.7−	40.9+	33.5+
23		23.6−	34.2+	22.6±	22.3−	22.6−	21.2−	27.0+	29.9+
$\bar{x}_{24}$		30.5	32.2	22.6	29.7	23.5	24.5	22.5	28.0

[a] Runs of 6 or more on each side of the median in a set of 24 are significant ($P = 0.05$). A run of 8 above or below the median in a set of 24 is significant ($P = 0.05$).
[b] + and − indicate values above and below the daily median speed; ± indicates values equal to the daily median speed.

firmed the presence of internal rhythmic control and showed the lability of the rhythm under the phasic control of light.

D. Constant Light

1. Swimming Speed

Further examination of the influence of internal control on daily rhythmicity continued with measurements of activity in constant light of low intensity. The rhythm could persist in the absence of a light cycle only if there were some degree of internal control, and this would support the evidence from the shifted-light experiments. We used constant light of low intensity in this experiment because total darkness would inhibit the ability of the fish to see, while constant light of higher intensity might contribute to a pathological condition.

Table VI. Hourly Median Speeds (cm/sec) Measured During Three Consecutive 7-hr Light Delays, Beginning on the Sixth Day of the Initial 7-hr Shift[a]

Hour		Day 1	2	3	4	5	6	7	8	9	10	11	12	13	14
0		26.8−[b]	28.4−	33.2−	42.9+	46.5+	49.3+	68.4+	49.3+	39.0+	29.9−	15.2−	33.8+	16.9−	18.2−
1	Light	23.4−	25.6−	37.2−	37.2−	45.3+	45.9+	57.8+	45.3+	34.9−	29.4−	27.5−	29.1−	18.2−	15.5−
2		21.9−	31.0−	44.1+	48.6+	47.9+	40.4+	37.6+	59.8+	27.9−	28.4−	29.6−	31.9+	20.4−	14.0−
3		25.7−	21.8−	37.6±	35.6−	41.9+	41.9+	47.9+	49.3+	46.5+	24.6−	45.3+	32.8+	25.6+	23.9−
4		20.1−	32.2+	37.6±	41.9+	40.9+	37.2+	45.9+	36.4+	50.8+	37.2+	57.8+	46.5+	41.4+	34.2+
5		23.6−	28.6−	25.0−	38.1+	41.9+	35.6−	30.5−	32.8+	40.9+	36.4+	59.8+	52.3+	38.1+	32.8+
6		39.4+	23.1−	19.3−	34.5−	39.9+	39.9+	37.2+	39.0+	46.5+	42.4+	46.5+	50.0+	50.8+	41.9+
7	Dark	35.3+	27.5−	30.5−	35.6−	41.9+	35.6−	47.9+	21.2−	33.5−	36.0+	52.3+	53.2+	47.2+	52.3+
8		28.9−	13.6−	25.0−	44.1+	39.0+	29.9−	35.3+	19.9−	31.9−	57.8+	63.2+	33.2+	35.3+	34.5+
9		29.9−	17.8−	12.8−	32.8−	34.2−	24.8−	32.2−	23.4−	26.8−	44.1+	44.7+	38.1+	37.2+	39.4+
10		31.0−	18.2−	25.4−	23.3−	31.9−	28.6−	29.4−	21.2−	30.2−	45.3+	50.0+	30.5+	42.4+	47.9+
11		33.5−	26.6−	22.3−	17.1−	30.5−	33.5−	24.5−	17.4−	38.1+	35.3+	28.6−	34.9+	49.3+	57.8+
12		26.2−	26.2−	28.9−	32.2−	30.2−	26.8−	33.2±	26.0−	25.7−	38.1+	36.4+	24.6−	54.0+	41.4+
13		19.0−	38.1+	39.0+	21.5−	32.8−	16.8−	28.6−	20.1−	53.2+	30.7−	40.9+	27.5−	38.1+	40.4+
14		42.9+	39.9+	42.9+	32.2−	37.2−	32.5−	33.2±	19.7−	48.6+	33.8+	30.5−	30.5+	25.4−	34.5+
15		47.2+	46.5+	62.0+	35.6−	34.2−	28.4−	29.6−	20.2−	60.9+	44.1+	34.9+	22.8−	30.5+	35.6+
16		41.9+	45.9+	41.9+	51.6+	21.8−	35.6−	32.8−	15.5−	50.8+	31.6+	31.0−	29.4−	16.2−	27.0−
17		41.9+	44.7+	41.4+	50.8+	22.3−	36.0±	32.2−	34.9+	35.3−	21.8−	33.5+	29.9−	14.7−	20.4−
18	Light	42.4+	47.2+	57.8+	47.2+	32.8−	39.0+	32.8−	31.0+	36.4+	16.2−	30.5−	27.9−	20.2−	26.6−
19		52.3+	45.3+	50.0+	50.8+	40.9+	44.1+	24.5−	24.6−	38.1+	13.5−	29.4−	21.0−	11.6−	23.9−
20		39.9+	45.3+	49.3+	59.8+	29.9−	38.5+	23.4−	27.5−	30.7−	15.5−	22.9−	20.2−	12.3−	18.2−
21		49.3+	45.3+	49.3+	33.2−	37.2−	36.0±	34.2+	32.2+	29.6−	18.4−	18.8−	18.6−	12.1−	19.3−
22		50.0+	38.1+	42.2+	44.7+	62.0+	41.4+	36.4+	34.9+	28.9−	28.9−	16.9−	23.3−	15.1−	19.5−
23		35.6+	37.6+	36.4−	41.4+	68.4+	54.0+	41.9+	41.9+	31.9−	16.4−	29.9−	21.5−	19.5−	19.0−
$\bar{x}_{24}$		34.4	31.6	37.6	37.7	38.1	36.0	33.2	29.3	35.9	31.2	32.3	30.2	25.5	29.9

[a] Runs of 6 or more on each side of the median in a set of 24 are significant ($P = 0.05$). A run of 8 above or below the median in a set of 24 is significant ($P = 0.05$).

[b] + and − indicate values above and below the daily median speed; ± indicates values equal to the daily median speed.

Following 2 days of a normal light regimen, we held the light intensity constant at 9 ft-candles beginning at morning light onset of the third day and continuing for 113 consecutive hours. Table VII shows the hourly median speeds obtained before and during constant light. At light onset of day 3, at the beginning of constant light and for about the next 42 hr, the fish showed a normal swimming pattern, i.e., high swimming speed during the hours of the natural day and low swimming speed during the hours of the natural night. Following this, on day 5. the third day of constant light, there was a partial breakdown of the rhythm which continued through the day. Starting at about hour 63 after constant light onset and

Table VII. Hourly Median Speeds (cm/sec) Measured for 55 hr under a 10.66-hr Photoperiod Followed by 113 hr under Constant Light Intensity of 9 ft-candles[a]

	Day						
Hour	1	2	3	4	5	6	7
0	8.4−[b]	10.5−	7.4−	9.4−	19.0−	10.2−	16.8−
1	4.4−	7.3−	6.7−	9.3−	29.4+	15.6−	14.4−
2	7.9−	7.4−	6.8−	10.1−	25.7−	11.3−	11.0−
3	8.0−	9.0−	7.4−	8.1−	29.4+	14.0−	12.7−
4	7.2−	8.7−	7.4−	9.5−	19.5−	9.9−	16.1−
5	7.4−	6.4−	5.8−	46.5+	23.9−	19.5−	16.6−
6	9.8−	9.6−	8.2−	25.7−	27.5+	19.5−	19.9−
Light onset			Constant light				
7	25.7+	22.6+	41.9+	26.4±	25.7−	19.7−	23.6−
8	30.7+	23.6+	37.2+	34.2+	34.2+	25.4+	24.3−
9	37.2+	24.5+	36.0+	30.5+	27.7+	25.4+	42.9+
10	32.5+	40.4+	39.9+	27.0+	25.4−	22.3+	29.9+
11	33.5+	27.7+	39.0+	27.7+	27.0+	27.0+	29.6+
12	31.0+	26.2+	30.7+	31.3+	36.4+	29.4+	32.8+
13	31.9+	30.5+	33.5+	41.4+	26.6+	32.2+	30.5+
14	22.1+	23.9+	28.9+	31.9+	31.0+	36.4+	31.6+
15	36.8+	26.6+	28.2+	28.4+	25.7−	39.0+	30.5+
16	33.5+	24.3+	27.9+	26.4±	31.0+	19.9−	24.6+
17	34.2+	22.3+	46.5+	21.8−	32.5+	25.2+	27.9+
Dark onset							
18	17.3+	11.8−	25.7+	29.6+	22.8−	22.6+	31.6+
19	10.6−	12.7+	22.0−	34.2+	24.1−	25.2+	31.0+
20	12.1−	12.3−	16.8−	24.8−	30.2+	21.8+	27.5+
21	11.0−	7.3−	12.1−	18.0−	8.8−	16.6−	19.0−
22	14.2−	9.3−	10.3−	11.6−	15.8−	15.8−	19.9−
23	8.6−	6.7−	15.2−	8.8−	9.9−	19.9−	23.6−
$\tilde{x}_{24}$	15.8	12.5	23.9	26.4	26.2	20.9	24.5

[a] Runs of 6 or more on each side of the median in a set of 24 are significant ($P = 0.05$). A run of 8 in a set of 24 above or below the median is significant ($P = 0.05$).
[b] + and − indicate values above and below the daily median speed; ± indicates values equal to the daily median speed.

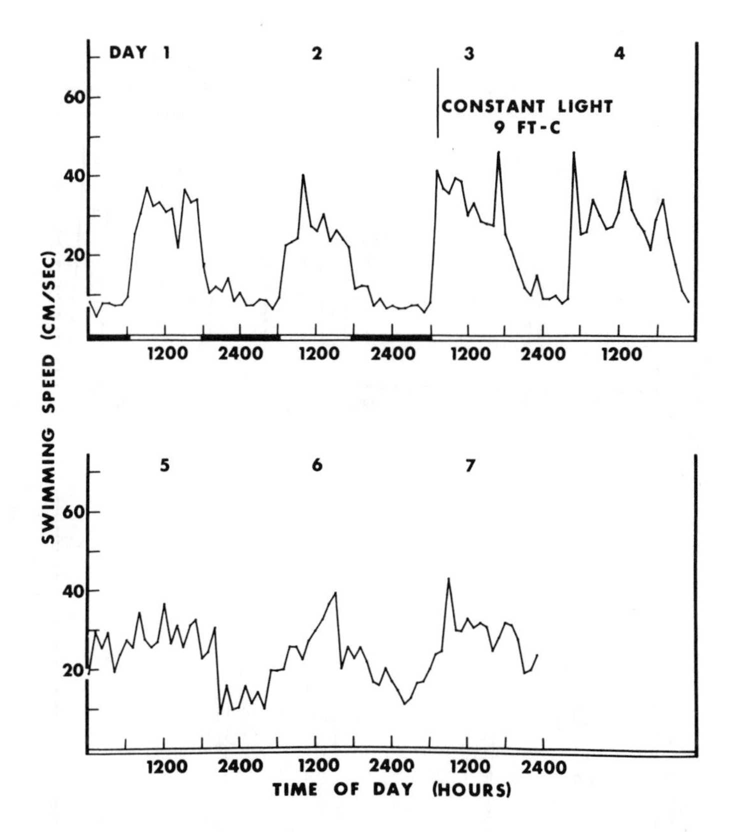

Fig. 9. Hourly median swimming speeds measured for 2 days under a 10.66-hr photoperiod followed by 113 hr of constant light of 9 ft-candles.

continuing for the next 51 hr, the rhythm was again well established, although it drifted forward 2–3 hr (Fig. 9). The persistence of a daily rhythm in the absence of continuing light cues was further confirmation for the presence of an internal component for daily rhythmicity.

2. Schooling

Measurements of schooling-group size recorded under constant light showed changes in schooling corresponding to the natural day–night cycle, suggesting internal control of fish-to-fish responsiveness. For about the first 41 hr following light onset on day 3, the pattern of schooling activity was similar to that of the preceding 2 days of normal light, i.e., strong schooling during the hours of the natural day (Fig. 10).

During day 5, the third day of constant light, the day–night school-

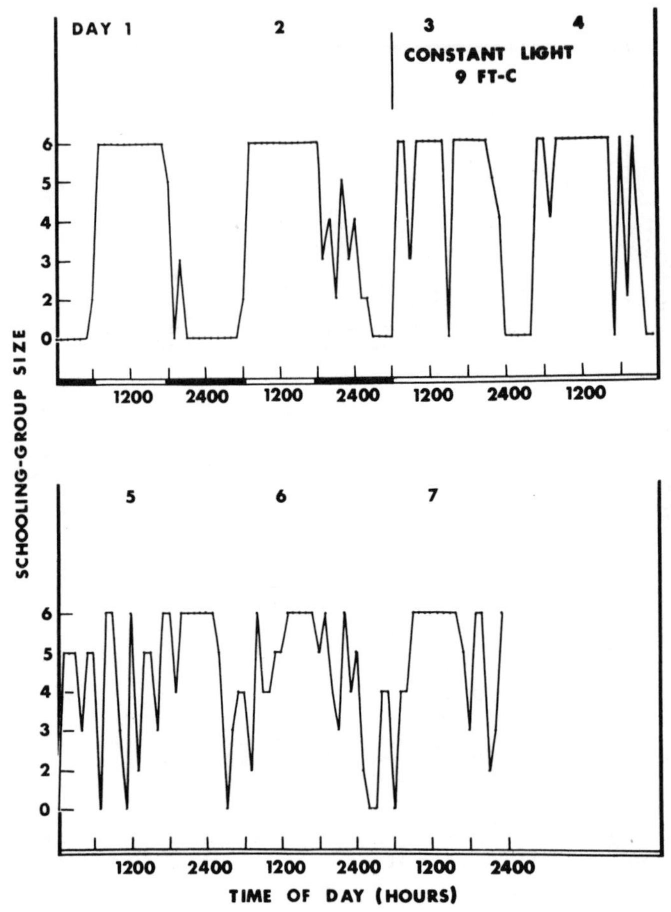

Fig. 10. Hourly medians of schooling-group size measured for 2 days under a 10.66-hr photoperiod followed by 113 hr of constant light of 9 ft-candles.

ing pattern broke down just as we had observed for the swimming-speed rhythm. From about 70 hr after constant light onset until the end of the observations, the day–night rhythm of schooling under constant light appeared to be somewhat re-established, although with a greater degree of variability in group size. The high frequency of small group size during the natural night and large groups during the natural day was closely correlated with the swimming-speed rhythm.

E. Retinal Rhythm Under Constant Darkness

If rhythmic activity in bluefish were phased by light and internally controlled, as our evidence indicated, we expected that there might be a related degree of internal control in the light–dark adaptation of the retina.

To investigate this, we measured the movement of the cones and pigment epithelium of young bluefish (14–17 cm) sampled periodically while the fish were kept in constant darkness (Olla and Marchioni, 1968).

Under a normal light–dark regimen, we had found that the retinal changes associated with light-and-dark adaptation followed the typical pattern in fishes for cones and pigment epithelium. Complete light adaptation took between 15 and 30 min, complete dark adaptation between 30 and 60 min. When the fish had become light-adapted, the cone ellipsoids lay close to the external limiting membrane and the pigment epithelium was expanded, remaining in that position until the light phase ended. When the fish had become dark-adapted, the cones lay close to the pigment epithelium, which was in a contracted condition. Thus under a regimen of alternating phases of light and dark, the cones and pigment epithelium moved from one extreme to the other with each change of phase. Under constant darkness, the cones continued to move relative to the natural photoperiod from the position normally associated with dark adaptation to that normally associated with light adaptation, but they did this gradually in a rhythmic pattern. But while the pigment epithelium had expanded and contracted with a change from light to darkness, it remained contracted throughout the whole period of constant darkness.

For comparative purposes, we plotted mean values of cone position against the swimming-speed rhythm under constant light (9 ft-candles) which was described earlier (Fig. 11). It was apparent from examination of the two curves that the dark-adapted position of the cone ellipsoids and low

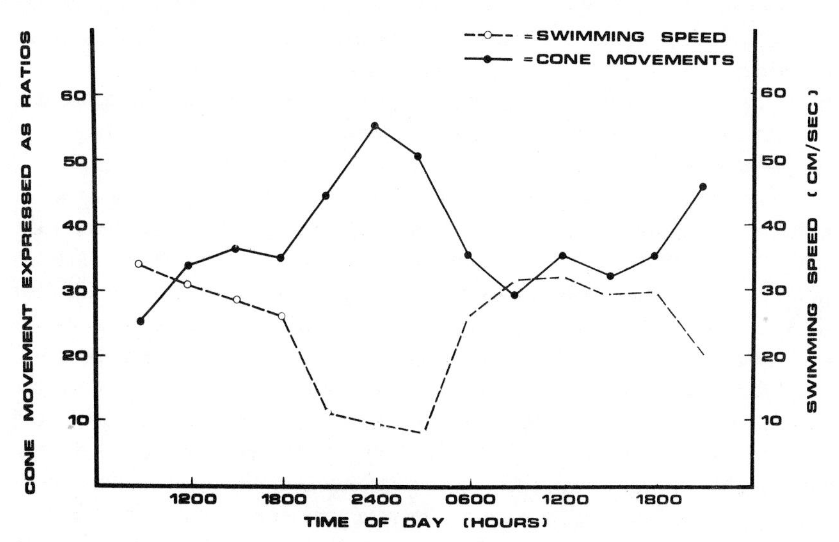

Fig. 11. Mean values of cone position from fish held in constant darkness compared with swimming activity under constant low illumination.

swimming activity of the bluefish held under constant low light occurred concurrently during the hours of the natural dark phase. Conversely, a light-adapted position of the cone ellipsoids and increased swimming activity occurred during the time of the natural light phase.

IV. DISCUSSION

It was clear from our study that day–night changes in the degree of activity were not solely related to changes in the daily light cycle. The day–night rhythm in swimming speed and group size persisted even in the absence of an externally imposed day–night cycle. Whether diurnal changes in swimming speed and group size are correlated because they depend on the same mechanism of internal control or whether one is dependent on the other does not alter the conclusion that day–night changes in responsiveness to species mates are, in part, internally controlled. During the day, individuals are more attracted to each other, forming schools that are significantly more cohesive than at night. Thines and Vandenbussche (1966) and more recently Shaw (1970) have postulated that day–night changes in schooling behavior may involve more than changing light levels. They have suggested that there may be some degree of internal fluctuation in responsiveness correlated with external environmental changes.

Day–night changes in responsiveness may relate to feeding in the bluefish. These fish are predators, using vision as a primary sense in feeding (Olla *et al.*, 1970). They would be most efficient in finding and capturing prey at light intensities permitting the highest visual acuity, i.e., during the day when they are also most active. Although there is a greater expenditure of energy during the day than during night, the chances that the fish could successfully ingest amounts of prey yielding energy equal to or greater than that expended would be greatest during the day.

The fact that cone movements which are related to light–dark adaptation appear to be under some degree of internal control may be related to the selective advantage of the rhythm (Olla and Marchioni, 1968). Through internal control the retina might be predisposed to the coming of light or dark, effectively lessening the time for adaptation. This would be a significant adaptation for a predator which is highly active during morning twilight.

Our experimental bluefish were of the same length as that part of the natural population that makes long northerly migrations in the spring and southerly migrations in the fall. By isolating our fish from changes in stimuli that might naturally influence these seasonal movements, we have been able to observe, in the aquarium, the role that one stimulus, day length, may play in these migrations. During photoperiods which corresponded to those occurring from spring to fall, the bluefish swam at significantly higher speeds than during the shorter (winter) photoperiods. These longer photoperiods included the times of spring and fall migration, spring spawning, and the summer, when fodder fish are most concentrated and abundant. Our observations at the shorter photoperiods, when speeds in the tank were lowest, corresponded to winter, when migration is ended and food is not readily abundant.

It must be pointed out that, although a strong correlation exists between photoperiod and activity, we have not demonstrated one way or the other whether the seasonal rhythm is entirely internally controlled or is cued at certain critical seasons by specific photoperiods or by some uncontrolled stimulus. It is our view that, as in the case of daily rhythmicity, seasonal changes in photoperiod act to synchronize the internal seasonal cycle of the individual and thus of the population as a whole.

ACKNOWLEDGMENTS

We wish to express our grateful appreciation to Enoch B. Ferrell for his advice on the statistical treatment of the data and to Lionel A. Walford for his criticism of the manuscript.

REFERENCES

Ali, M. A., 1961, Histophysiological studies on the juvenile Atlantic salmon (*Salmo salar*) retina. II. Responses to light intensities, wavelengths, temperatures, and continuous light or dark, *Canad. J. Zool.* **39:** 511.

Arey, L. B., and Mundt, G. H., 1941, A persistent diurnal rhythm in visual cones, *Anat. Rec.* **79** (Suppl.): 5 (abst.).

Aschoff, J., 1965, ed., "Circadian Clocks," Proc. Feldafing Summer School 1964, North Holland Publishing Co., Amsterdam.

Baggerman, B., 1957, An experimental study on the timing of breeding and migration in the three-spined stickleback (*Gasterosteus aculeatus* L.), *Archiv. Néerl. Zool.* **12(2):** 105.

Baggerman, B., 1959, The role of external factors and hormones in migration of sticklebacks and juvenile salmon, *in* "Comparative Endocrinology" (A. Gorbman, ed.) pp. 24–37, John Wiley and Sons, Inc., New York.

Baggerman, B., 1962, Some endocrine aspects of fish migration, *Gen. Comp. Endocrinol. Suppl.* **1:** 188.

Bünning, E., 1964, "The Physiological Clock; Endogenous Diurnal Rhythms and Biological Chronometry," Springer-Verlag, Berlin.

Cloudsley-Thompson, J. L., 1961, "Rhythmic Activity in Animal Physiology and Behaviour," Academic Press, New York.

Cold Spring Harbor Symposium on Biological Clocks, 1960, *Cold Spring Harbor Symp. Quant. Biol.* **25.**

de Groot, S. J., 1964, Diurnal activity and feeding habits of plaice, *Rapp. Proc. Verb. Cons. Internat. Explor. Mer* **155:** 48.

Ferrell, E. B., 1958, Control charts for Log-Normal universes, *Industrial Quality Control* **15(2):** 4.

Harker, J. E., 1964, "The Physiology of Diurnal Rhythms," University Press, Cambridge.

John, K. R., and Gring, D. M., 1968, Retinomotor rhythms in the bluegill, *Lepomis macrochirus*, *J. Fish. Res. Bd. Canad.* **25:** 373.

John, K. R., and Haut, M., 1964, Retinomotor cycles and correlated behavior in the teleost *Astyanax mexicanus* (Fillipi), *J. Fish. Res. Bd. Canad.* **21:** 591.

John, K. R., and Kaminester, L. H., 1969, Further studies on retinomotor rhythms in the teleost *Astyanax mexicanus*, *Physiol. Zool.* **42(1):** 60.

John, K. R., Segall, M., and Zawatzky, L., 1967, Retinomotor rhythms in the goldfish, *Carassius auratus*, *Biol. Bull.* **132:** 200.

Kruuk, H., 1963, Diurnal periodicity in the activity of the common sole, *Solea vulgaris* Quensel, *Neth. J. Sea Res.* **2:** 1.

Liley, N. R., 1969, Hormones and reproductive behavior in fishes, *in* "Fish Physiology" (W. S. Hoar and D. J. Randall, eds.) Vol. 3, pp. 73–116, Academic Press, New York.

Olla, B. L., and Marchioni, W. W., 1968, Rhythmic movements of cones in the retina of bluefish, *Pomatomus saltatrix*, held in constant darkness, *Biol. Bull.* **135:** 530.

Olla, B. L., Marchioni, W. W., and Katz, H. M., 1967, A large experimental aquarium system for marine pelagic fishes, *Trans. Amer. Fish. Soc.* **96:** 143.

Olla, B. L., Katz, H. M., and Studholme, A. L., 1970, Prey capture and feeding motivation in the bluefish, *Pomatomus saltatrix*, *Copeia* **1970(2):** 360.

Olmstead, P. S., 1958, Runs determined in a sample by an arbitrary cut, *Bell System Tech. J.* **Jan. 1958:** 55.

Shaw, E., 1970, Schooling in fishes: Critique and review, *in* "The Development and Evolution of Behavior" (L. R. Aronson, E. Tobach, J. S. Rosenblatt, and D. S. Lehrman, eds.) pp. 452–480, W. H. Freeman and Co., San Francisco.

Thines, G., and Vandenbussche, E., 1966, The effects of alarm substance on the schooling behaviour of *Rasbora heteromorpha* Duncker in day and night conditions, *Anim. Behav.* **14:** 296.

Tukey, J. W., 1959, A quick, compact, two-sample test to Duckworth's specifications, *Technometrics* **1:** 31.

Welsh, J. H., and Osborn, C. M., 1937, Diurnal changes in the retina of the catfish, *Ameiurus nebulosus*, *J. Comp. Neurol.* **66:** 349.

Wigger, H., 1941, Diskontinuität und Tagesrhythmik in der dunkelwanderung retinaler Elemente, *Z. Vergl. Physiologie* **28:** 421.

Winn, H. E., Salmon, M., and Roberts, N., 1964, Sun-compass orientation by parrot fishes, *Z. Tierpsychol.* **21(7):** 798.

Woodhead, A. D., and Woodhead, P. M. J., 1965, Seasonal changes in the physiology of the Barents Sea cod, *Gadus morhua* L., in relation to its environment: I. Endocrine changes particularly affecting migration and maturation, *Spec. Publ. Internat. Comm. NW Atlant. Fish.* **6:** 691.

Woodhead, P. M. J., 1966, The behaviour of fish in relation to light in the sea, *in* "Oceanography and Marine Biology" (H. Barnes, ed.) Vol. 4, pp. 337–403, George Allen and Unwin, Ltd., London.

Chapter 9

BEHAVIOR OF SYMBIOTIC FISHES AND SEA ANEMONES

Richard N. Mariscal

Department of Biological Science
Florida State University
Tallahassee, Florida

I. INTRODUCTION

Due to a general misunderstanding of de Bary's (1879) original definition, there has been much confusion during the past century regarding the term "symbiosis" and its subcategories. Fortunately, this situation now seems to have been corrected. Many texts and standard reference works, such as that edited by Henry (1966), now correctly consider symbiosis simply to mean a "living together," as originally defined by de Bary. Under this all-inclusive heading are grouped the three major subdivisions of "mutualism," "commensalism," and "parasitism "(Allee *et al.*, 1949; Davenport, 1955; Noble and Noble, 1964; Henry, 1966).

Mutualism is a symbiotic association in which both partners benefit; commensalism is an association in which one partner benefits and the other is neither harmed nor benefited; parasitism is a situation in which one partner benefits and the other is harmed. Other categories and subcategories generally can be grouped with these three and will not be considered here.

This chapter deals with the behavioral interactions of one of the better-known, but lesser-studied, symbiotic associations: that of sea anemones and pomacentrid fishes from the tropical Indo-Pacific (Fig. 1). This association is probably best thought of as mutualistic, rather than commensal as it is more often described (Mariscal, 1970*c*).

Fig. 1. Laboratory electronic flash photograph of two juvenile *Amphiprion xanthurus* specimens over the oral disc of a *Stoichactis* sea anemone. When associating with a short-tentacled anemone such as this one, *Amphiprion* specimens often "bathe" among the folds of the oral disc as they do in the longer tentacles of a related species such as *Radianthus*.

Although isolated examples exist of fishes associating with coelenterates, the members of the genera *Amphiprion* (about 16 species) and *Premnas* (one species) appear to be the main ones which commonly, and perhaps exclusively, live in this fashion throughout life. *Premnas biaculeatus* (shown in the color frontispiece of Henry, 1966, and in Mariscal, 1970*b*) is very similar in gross morphology, color pattern, ecology, and behavior to the various species of *Amphiprion*. It is apparently found with sea anemones both as a juvenile and an adult. Other members of the family Pomacentridae, such as *Dascyllus trimaculatus* and *Dascyllus albisella*, commonly associate with sea anemones as juveniles, but not as adults.

Fishes are generally considered to be natural prey of sea anemones, since they may be captured and killed by the venomous nematocysts covering the anemone's tentacles (Gudger, 1941; Mariscal, 1966*a*). In spite of this, the various species of *Amphiprion* live in intimate association with the anemone, frequently swimming through and violently battering the anemone's tentacles with their bodies. A detailed discussion of the protective mechanisms of anemone fishes and related topics has been given elsewhere (Mariscal, 1966*a,b*, 1967, 1969, 1970*a*, 1971).

II. THE ANEMONES

A. General Description

Only a few of the many reports concerning the anemone–fish relationship give enough information to permit specific identification of the partners. Although *Stoichactis* (Fig. 1) is the only genus of sea anemones commonly mentioned as the host for symbiotic fish, some 11 species of sea anemones from six genera and four families are known to harbor *Amphiprion* or *Premnas* (Table 1). In collecting these anemones and fishes throughout the Indo-Pacific, I found that *Radianthus ritteri* was probably the most common species containing *Amphiprion*. Next in abundance were the various species of *Stoichactis*. Since both *Radianthus* and *Stoichactis* are members of the family Stoichactiidae, it is probably safe to say that this family (containing at least two genera and six species of known symbiotic anemones) is the one primarily involved in this symbiosis throughout the Indo-Pacific and Red Sea. The family Actiniidae containing two genera (*Physobrachia* and *Macrodactyla*), the family Thalassianthidae with one genus (*Cryptodendrum*), and supposedly the family Actinostolidae with one genus (*Parasicyonis*) are the other anemone taxa involved.

The stoichactiid sea anemones are the largest in the world. In an island group off western Thailand, I measured and photographed a *Stoichactis giganteum* underwater on two separate days and found it to be 1.24 m in diameter. This appears to be the largest sea anemone ever accurately measured and recorded. This anemone contained two large *Amphiprion xanthurus* individuals (Mariscal, 1970*b*). The smallest symbiotic sea anemone found was a specimen of *Radianthus ritteri* in the same group of islands. This anemone measured about 10 cm in diameter and contained a single juvenile *Amphiprion akallopisos*.

The symbiotic anemones exhibit a wide array of tentacle and column coloration, both intra- and interspecifically. Within the same species (e.g., *R. ritteri*), I have collected anemones whose tentacles were either a drab brown, orange, or green and whose column coloration was either brown, orange, deep red, or bright purple. Usually, but not always, anemones of a more-or-less similar coloration were found in the same general reef area. Collingwood (1868*a*,*b*), Saville-Kent (1893, 1897), Yonge (1930), and Stephenson (1946) have reported similar observations regarding color variation among these anemones.

B. Ecology

The symbiotic sea anemones were most commonly found in shallow reef areas protected from heavy wave action. Protected bays or lagoons

Table I. World-Wide Summary of Reported Associations of *Amphiprion* and *Premnas* with Sea Anemones[a]

No.	Fish species	Anemone species	Location	Reference
1.	*Amphiprion xanthurus*	1. *Stoichactis kenti*	Takanupe I., Solomon Is.	Mariscal (1966*a*)
		2. *Stoichactis giganteum*	Bougainville I., Solomon Is.	Mariscal (1966*a*)
			Funidu I., Malé Atoll, Maldive Is.	Mariscal (1966*b*)
			Pipidon I., W. Thailand	Mariscal (1966*b*)
			Pipilek I., W. Thailand	Mariscal (1966*b*)
		(This is possibly anemone 2 of Verwey.)	Batavia, Indonesia	Verwey (1930)
		(This is probably the *Discosoma* sp. reported.)	Maldive and Nicobar Is.	Eibl-Eibesfeldt (1960, 1965)
		3. *Radianthus ritteri*[b]	Funidu I., Malé Atoll, Maldive Is.	Mariscal (1966*b*)
		(This is probably anemone 3 of Verwey.)	Batavia, Indonesia	Verwey (1930)
		4. *Radianthus kuekenthali*	New Britain I., Territory of New Guinea	Mariscal (1966*b*)
			Nicobar Is.	Eibl-Eibesfeldt (1960, 1965)
		5. *Physobrachia ramsayi*	Funidu I., Malé Atoll, Maldive Is.	Mariscal (1966*b*)
		6. *Cryptodendrum adhesivum*	Hulele I., Malé Atoll, Maldive Is.	Mariscal (1966*b*)
		7. *Radianthus koseirensis*	Hulele I., Malé Atoll, Maldive Is.	Mariscal (1966*b*)
		8. *Macrodactyla gelam*	Moreton Bay, N.S.W., Australia	C. Hand (personal communication)
		9. *Stoichactis haddoni* (This may be identical to *Stoichactis giganteum*.)	W. Australia (Shark's Bay)	Saville-Kent (1897)
2.	*Amphiprion percula*	1. *Stoichactis kenti*	Bougainville I., Solomon Is.	Mariscal (1966*a*)
			New Britain I., Territory of New Guinea	Mariscal (1966*b*)
		(This is probably anemone 1 of Verwey.)	Batavia, Indonesia	Verwey (1930)
		S. kenti	Queensland, Australia	Saville-Kent (1893)
			Low Isles, Queensland, Australia	Yonge (1930)
			Luzon, Philippines	Davenport and Norris (1958)

	2. *Radianthus ritteri*	New Britain I., Territory of New Guinea	Mariscal (1966*b*)
		Vanikoro I., Santa Cruz Is.	Mariscal (1966*b*)
		Pipilek I., W. Thailand	Mariscal (1966*b*)
		Pipidon I., W. Thailand	Mariscal (1966*b*)
		Nicobar Is.	Eibl-Eibesfeldt (1960, 1965)
	(This is probably anemone 4 of Verwey.)	Batavia, Indonesia	Verwey (1930)
3. *Amphiprion akallopisos*	1. *Radianthus ritteri*	Jadini, Kenya, E. Africa	Mariscal (1966*b*)
		Mahé I., Seychelles Is.	Mariscal (1966*b*)
		Pipilek I., W. Thailand	Mariscal (1966*b*)
		Pipidon I., W. Thailand	Mariscal (1966*b*)
		Nicobar Is.	Eibl-Eibesfeldt (1960, 1965, 1967)
	(This is probably Anemone 4 of Verwey.)	Batavia, Indonesia	Verwey (1930)
4. *Amphiprion perideraion*	1. *Stoichactis kenti*	New Britain I., Territory of New Guinea	Mariscal (1966*a*)
	2. *Radianthus ritteri*	Bougainville I., Solomon Is.	Mariscal (1966*a*)
	3. *Parasicyonis actinostoloides*	Amami Is., Japan	Okuno (1963), Araga (1964)
5. *Amphiprion melanopus*	1. *Physobrachia ramsayi*	Tutuila I., E. Samoa	Mariscal (1966*a*, *b*)
	2. *Radianthus kuekenthali*	New Britain I., Territory of New Guinea	Mariscal (1966*b*)
6. *Amphiprion nigripes*	1. *Radianthus ritteri*	Funidu I., Malé Atoll, Maldive Is.	Mariscal (1966*b*)
		Hulele I., Malé Atoll, Maldive Is.	Mariscal (1966*b*)
		Maldive Is.	Eibl-Eibesfeldt (1965, 1967)
(Misidentified as *A. melanopus* in Fig. 9, 1965, and Fig. 8, 1967.)			
7. *Amphiprion bicinctus*	1. *Radianthus* sp.	Jadini, Kenya, E. Africa	Mariscal (1966*b*)
	2. *Stoichactis giganteum*	Ghardaqa, Red Sea	Gohar (1948)
	S. giganteum (?)	Eilat, Red Sea	Graefe (1964)
	S. giganteum (?)	Eilat, Red Sea	Fishelson (1965)
	3. *Stoichactis haddoni*	Queensland, Australia	Saville-Kent (1893)
	4. *Radianthus koseirensis*	Eilat, Red Sea	Graefe (1964)
	(= *Antheopsis koseirensis*)	Eilat, Red Sea	Fishelson (1965)

Table I (Cont'd)

No.	Fish species	Anemone species	Location	Reference
		5. *Physobrachia quadricolor* (= *P. ramsayi*?)	Ghardaqa, Red Sea	Gohar (1948)
8.	*Amphiprion ephippium*	1. *Physobrachia ramsayi*	Pipilek I., W. Thailand	Mariscal (1966*b*)
			Pipidon I., W. Thailand	Mariscal (1966*b*)
			Vanikoro I., Santa Cruz Is.	Mariscal (1966*a*)
			W. Australia (Lacepede Is.)	Saville-Kent (1897)
		(This probably anemones 5 and 6 of Verwey.)	Batavia, Indonesia	Verwey (1930)
9.	*Premnas biaculeatus*	1. *Physobrachia ramsayi* (Anemones 5 and 6 of Verwey?)	Batavia, Indonesia	Verwey (1930)
	P. biaculeatus (?)		W. Australia (Lacepede Is.)	Saville-Kent (1897)
(See Plate XXXIX for fish—this could also be a dark morph of *A. percula*, although it is impossible to say from the information available.)				
	P. biaculeatus (?)	2. *Stoichactis kenti*	W. Australia (Lacepede Is.)	Saville-Kent (1897)
		(Plate XXXIX—not *S. haddoni* as listed in caption.)		

[a] In general, only those cases are cited where enough information is available for a probable identification of the anemones and fishes.
[b] *R. ritteri* is not the "normal" host anemone for this fish, and only a single juvenile *A. xanthurus* was found sharing this anemone with *A. nigripes* in the Maldives.

inside fringing or barrier reefs were common habitats. Surprisingly, the stoichactiid anemones seemed more common in the relatively dead portions of these reef areas rather than among the actively growing corals and alcyonarians (Mariscal, 1969).

Four main types of habitat were utilized by the symbiotic anemones, depending on the species. For example, *Radianthus ritteri* was normally found on relatively exposed dead portions of small coral boulders or ledges. Isolated anemones with fish were common, but this species also formed clusters of four or five individuals among which the fish set up subterritories (Bowman and Mariscal, 1968). At the other extreme was *Physobrachia ramsayi*, which often formed fairly large, but obscure, aggregations in crevices and holes in dead coral boulders (Mariscal, 1966*a*). Somewhere between these two extremes was *Stoichactis giganteum*, which extended its very large oral disc out over the reef surface but often kept its pedal disc anchored deep in a pile of coral rubble or a crevice which had to be carefully broken open to collect the anemone (Mariscal, 1966*a*). Other species such as *Stoichactis kenti* might be found on exposed sand flats with the pedal disc anchored to piece of coral rubble deep beneath the surface of the sand.

Some workers (e.g., Caspers, 1939) have doubted that the symbiotic anemones possessed nematocysts, while others have believed that these intracellular organelles were present, but incapable of aiding in prey capture (Buhk, 1939). My light and electron microscope observations, as well as those of others, reveal that the tentacles of the symbiotic anemones definitely contain numerous nematocysts and that they are fully functional and capable of capturing fishes as well as other prey (Gohar, 1934, 1948; Herre, 1936; Koenig, 1960; Eibl-Eibesfeldt, 1960). I have also observed fish capture by stoichactiids in shipboard aquaria. Other food of anemones includes mollusks, worms, and crustaceans.

C. Behavior

1. Locomotion

Although sea anemones are generally considered to be sedentary, they can crawl relatively rapidly when detached. When anemones are placed in shipboard aquaria after being detached from their normal substrate, a prolonged period of movement that might be termed "search behavior" occurs until the anemone finds an appropriate substrate for attachment. *Physobrachia ramsayi*, normally a crevice-dwelling form, moved at a rate of 20 cm/hr and halted only when it found conditions approximating those in the field. *Stoichactis kenti* moved at a rate of about 18 cm/hr. In both cases, the anemones' symbiotic fishes remained with them.

2. Reaction to Environment Stimuli

Tropical symbiotic anemones respond to light. In the absence of direct daylight in the laboratory, the anemones often become flaccid. However, if the anemones are suddenly uncovered in the direct sunlight, their tentacles become turgid and begin waving about. Covered control anemones do not display this reaction. When covered or moved into a darkened laboratory, the anemones' tentacles again lose their turgidity. Although the receptors for this behavior are unknown and its function unclear, it may somehow be associated with the presence of symbiotic algae in some anemones' tissues which have been shown to be involved in their nutrition (Smith *et al.*, 1969).

Tactile stimulation of a symbiotic anemone's tentacles by either inorganic or organic material results in the tentacles bending toward the site of contact. This is often followed by contraction of the tentacles and oral disc on that side.

3. Reaction to Contact with Anemone Fishes

a. Contact with an Anemone's Own Fish. Tactile stimulation by the anemone's own symbiotic fish generally elicits no response on the part of the anemone unless the fish rather violently batters the tentacles or oral disc. Even this results in only a slight contraction and immediate expansion. Some workers have found that anemones respond to a tactile "massaging" of the column or oral disc area. For example, Gohar (1948), Verwey (1930), and Herre (1936) reported how contracted anemones expanded rapidly upon being touched by their symbiotic fishes. Although controlled observations are difficult, I have observed similar occurrences.

More recently, Ross and Sutton (1968) have demonstrated that the symbiotic anemone *Calliactis* (which forms associations with hermit crabs) will relax and detach its pedal disc from the substrate, due either to tactile stimulation by the crab or artificial stimulation by the observer with pipecleaners or electrical stimuli. I have conducted similar experiments with *Calliactis tricolor* using a camel's hair brush, cotton-tipped swabs, or simply the fingers. A contracted anemone which is gently stroked around the pedal disc may completely relax and expand within a minute or so and then slowly detach its pedal disc following several minutes of continuous stimulation.

b. Contact with Isolated Anemone Fishes. Anemone fishes which have been isolated from any contact with their sea anemones for 24 hr or longer, and which are consequently unprotected, elicit an entirely different reaction from the same sea anemones, whether with or without fishes. In this case, the anemone's tentacles adhere to the fish, contract sharply, and bend toward the mouth in a typical prey-capture response. Occasionally, a wave of tentacle contraction sweeps around the oral disc from the point of contact by the fish. Clearly, the isolated fishes now lack something which formerly protected them. This appears to be some factor associated with the mucous coat of the fishes (Mariscal, 1970*a*). Isolation of the sea anemones has no effect on the results.

III. THE FISHES

A. General Description

Regardless of where found, the size of the various species of *Amphiprion* and *Premnas* in the field generally range from 2 to 5 inches in total lengh. Nearly all have a basic body color of some shade of orange, red, yellow, brown, or black which is broken up by one to three vertical white bars of varying width. Some lack bars entirely, while one species has a single dorsal white stripe running the length of the body. Some have vertical white bars as juveniles but lose these as adults, while others undergo a marked color change as they mature, occasionally in conjunction with a loss of the vertical bars. These latter changes have led, not surprisingly, to a good deal of taxonomic confusion which in some cases can be clarified only by raising the juveniles in the laboratory or by collecting a range of age classes in a specific locality. Schultz (1953, 1966), Schultz *et al.* (1960), and Mariscal (1966*b*) give good illustrations of all the members of the genus *Amphiprion*, as well as *Premnas biaculeatus*.

B. Ecology

It seems likely that all species of *Amphiprion* and probably the single species of *Premnas* are to be found associated with sea anemones in the field. In observing hundreds of specimens of *Amphiprion* belonging to 11 species throughout the Indo-Pacific, I have never seen one without its symbiotic anemone somewhere in the vicinity, into which it retreated when pursued. Experiments in which *Amphiprion* individuals are separated from their anemones in the field (Eibl-Eibesfeldt, 1960; Mariscal, 1966*b*) or the anemones removed from their aquaria (Sluiter, 1888; Verwey, 1930; Coates, 1964) show that in the presence of predatory fishes, some anemone fishes are quickly eaten.

As previously noted, when an anemone moves, its symbiotic fish move with it. This is interesting in view of Abel's (1960*c*) and my observations of various species of *Amphiprion* returning to the exact place where either their eggs or anemone formerly had been. For example, if one removes the anemone of *Amphiprion percula* from its normal field position on a coral boulder, the fish will return to the exact spot from which the anemone was taken, even though the anemone may have been set down only a foot or so away. After some searching and hovering, the fish will discover the anemone nearby and then move to it. Even then, the fish frequently continues to make forays back to the anemone's former location. Moving an anemone from one part of a generally featureless aquarium to another results in the fish staying with or finding the anemone rather quickly. However, for several days the fish may continue to return to the exact spot from which the anemone has been removed. This behavior finally wanes and the fish takes up a normal existence with its anemone, no longer returning to any particular part of the aquarium. These, as well as other observations of *Amphiprion*, indicate a well-developed sense of spatial orientation which is not overcome by any single environmental feature such as the presence of the anemone in a new location. Rather, the fishes seem to recognize and remember the general features of their environment, the sum total of which may be more important than any one of them. This indicates a fairly sophisticated evaluation of sensory input worthy of detailed investigation.

There are reports of from one to ten *Amphiprion* individuals being found in one sea anemone. The number appears to be species specific, since for some species such as *Amphiprion xanthurus* usually only a pair of adults is found with a single anemone. On the other hand, I have always found four to eight *Amphiprion percula* individuals of all age classes inhabiting a single anemone. With others, such as *Amphiprion akallopisos*, *A. nigripes*, and *A. ephippium*, five or six fish commonly associate with a cluster of anemones of the same species and move freely from one to the other (Bowman and Mariscal, 1968; Mariscal, 1966*b*, 1970*b*).

C. Behavior

1. Behavior of Fishes Living with Anemones

a. Acclimation Behavior. The acclimation of anemone fishes to new anemones is probably the most pertinent behavioral interaction of this symbiosis.

At the start of acclimation behavior, an anemone fish which has been isolated from anemones for approximately 24 hr will be stung upon first contact with the anemone's tentacles; at the end of acclimation (lasting from 10 min to several hours, depending on the anemone species), the same fish will no longer be stung by the same anemone. Obviously, some change either in the fish or the anemone during this period results in protecting the fish.

Although Whitley (1932) and Gohar (1948) had earlier observed this, it remained for Davenport and Norris (1958) to make the first thorough study of the acclimation of *Amphiprion*. More recently, Mariscal (1965, 1966*a*, 1969, 1970*a*, 1971) has studied this in some detail, while Koenig

(1960), Oesman (1961), and Graefe (1963, 1964) have all observed it. Stevenson (1963) has observed the same behavior for *Dascyllus albisella*, while Mariscal (1966*b*) has observed it for *D. albisella* and *D. trimaculatus*. Blösch (1961, 1965) and Fishelson (1965) have also commented on acclimation but appear to have underestimated its significance.

Although a good deal of individual variation exists in the various elements of acclimation behavior, it is usually initiated by a fish cautiously nosing or nibbling a tentacle or clump of tentacles (Fig. 2). This may be repeated and is often interspersed with periods of hovering above or just to one side of the anemone. The fish then moves in to touch the tentacles gently with its pelvic, anal, or caudal fins, each contact of which causes the same initial response: a strong adhesion of the tentacle to the fish followed by a sharp jerking back by the fish and subsequent curling away and contraction of the tentacle. The fish gradually begins increasing the degree of contact and penetration of the anemone's tentacles, returning almost immediately after being stung to begin the same process over again. Although most fish move forward to contact the anemone's tentacles anteriorly, some hover above the anemone's disc and settle tail-down into the tentacles. Once on the oral disc, the fish backs tail-first into the tentacles and, upon

Fig. 2. Close-up laboratory electronic fish photograph of an *Amphiprion xanthurus* in the process of acclimating to a *Stoichactis kenti* sea anemone. Note the nibbling of the tentacles, a common behavioral act associated with acclimation.

being stung, rapidly swims up and out of the tentacles, only to return and repeat the behavior. As the tentacle penetration increases, there is a concomitant decrease in tentacle adhesion and contraction, indicating that the fish is becoming partially protected from the anemone's nematocysts. Finally, the fish begins a more general penetration and bathing among the tentacles with little or no response on the part of the anemone, indicating acclimation is complete,

The time of acclimation varies with the species of anemone. *Amphiprion xanthurus* acclimated to the tropical anemone *Stoichactis kenti* in about 10 min, to the California anemone *Anthopleura xanthogrammica* in about 1 hr, and to *Anthopleura elegantissima* (California) in about 45 hr in the one case where acclimation was completed (Mariscal 1970*a*).

Two different hypotheses attempt to explain what is occurring during acclimation. One suggests a change in the anemone during acclimation due to behavioral or physico-chemical stimuli from the fish (Gohar, 1948; Hackinger, 1959; Koenig, 1960; Graefe, 1963, 1964; Blösch 1961, 1965). This implies that the anemone is somehow controlling its nematocyst discharge in response to stimuli received from the fish. Only Graefe and Blösch of this group of authors have performed any experiments in this regard. Unfortunately, the results of these experiments lend themselves to several alternative hypotheses due to the lack of necessary controls (see Mariscal, 1969, 1970*a*, 1971).

The second hypothesis suggests that the anemone fish is undergoing a change during acclimation, apparently associated with the mucous coat, which results in its protection (Davenport and Norris, 1958; Eibl-Eibesfeldt, 1960, 1967; Mariscal, 1965, 1967, 1969, 1970*a*, 1971; Schlichter, 1967, 1968). The implication here is that the anemone, even with fish, is a passive partner fully capable of stinging unprotected fish. Abel (1960*b*) found that the mucus of *Gobius bucchichii* was responsible for its protection from a Mediterranean anemone, but he did not observe acclimation behavior. Fishelson (1965), working with *Amphiprion bicinctus* in the Red Sea, found no evidence for acclimation behavior and concluded that the Red Sea symbiotic anemones "are not adapted to the catching of living fish, and that this is the basic condition for the association of fish with them." Significant, however, is the fact that Fishelson tested *A. bicinctus* specimens which were already living with anemones against new anemones of the same species, and one would not expect apparently acclimated fish to undergo acclimation again. Schlicter (1967) has since verified in field experiments that Red Sea *A. bicintus* does indeed require a period of acclimation before it is protected from the local anemones.

Due to the lack of necessary controls, a number of confusing misconceptions concerning the protection of anemone fishes have been introduced into the literature in recent years. Since these misconceptions center on the acclimation process, perhaps it would be well to review again the rationale for experiments regarding this phenomenon. If an acclimated anemone fish is *not* stung either by anemones with acclimated fish or those without them, this suggests that some change has occurred in the fish during acclimation. If, on the other hand, an anemone containing acclimated fish at the time of the experiment does *not* sting unacclimated fish of the same species, this would suggest that some change has occurred in the anemone. This means that reciprocal experiments using acclimated and unacclimated fish of the same species tested against the same species of sea anemone both with and without acclimated fish are essential for any meaningful understanding of the protective mechanisms. If more than one species of sea anemone is used in the same experiments, it is even more critical to run careful reciprocal experiments with the various species in order to evaluate properly the results.

Table II shows a series of such reciprocal experiments. It will be seen that in all cases only the unacclimated fish were stung by the three species of sea anemones, regardless of whether or not the anemones had acclimated fish living with them at the time of the experiment. Similarly, none of the acclimated fish were stung by any of the anemones, either with or without the acclimated fish. These results show that only the condition of the fish and not the condition of the sea anemone is important in protecting the symbiotic fish, Experiments using different species of sea anemones show that an *Amphiprion* protected from one species of sea anemone is not automatically protected from other species (Mariscal, 1970*a*). For example, an *Amphiprion xanthurus* acclimated to *Anthopleura xanthogrammica* was not protected from *Anthopleura elegantissima*, a very close relative, which requred a longer time of acclimation. Likewise, a fish acclimated to *Stoichactis kenti* was not protected from *Anthopleura xanthogrammica*, also requiring a longer time of acclimation.

Experiments by Davenport and Norris (1958) and Mariscal (1966*b*, 1970*a*, 1971) demonstrated that the mucous coating of an acclimated *Amphiprion* is responsible for its protection. If this mucous is carefully removed, the acclimated or partially acclimated fish becomes immediately deacclimated and is stung upon every contact with the tentacles of its former anemone.

b. Territoriality. *Amphiprion* territorial behavior has been known for many years (Horst, 1903; Verwey, 1930; Moser, 1931; Gohar, 1934, 1948; Herre, 1936; Ladiges, 1939; Eibl-Eibesfeldt, 1960; Koenig 1960; Oesman, 1961; Schneider, 1964; Mariscal, 1966*b*, 1970*b*). Using colored models, Fricke (1966) has evoked intraspecific aggression and territorial defense from *Amphiprion bicinctus* in the field. Aronson (1957) discusses this subject as it relates to reproductive behavior.

Table II. Experiments Presenting *Amphiprion xanthurus* (Acclimated and Unacclimated) to Three Species of Sea Anemones, With and Without Fish[a]

Condition of fish	Condition of sea anemone							
	Stoichactis kenti		*Anthopleura xanthogrammica*		*Anthopleura elegantissima*			
Amphiprion xanthurus	Without fish	With fish	Without fish	With fish	Without fish	With fish	Total number of experiments	Summary of results
Unacclimated	+ (9)	+ (8)	+ (9)	+ (9)	+ (4)	+ (1)	40	Unacclimated fish all stung
Acclimated	— (6)	— (7)	— (8)	— (10)	— (2)	— (1)	34	Acclimated fish not stung
Total number of experiments	15	15	17	19	6	2	74	

[a] In all cases, the anemones with fish contained only the same species as that being tested. The numbers in parentheses refer to the number of experiments run (+ = fish stung, — = fish not stung, 0 = no data).

Amphiprion territorial defense is not strong in the absence of a sea anemone. Although isolated *Amphiprion xanthurus* individuals set up territories in the vicinity of holes, corners, and objects in the aquarium, these are only weakly defended. However, after an anemone has been present for several minutes, attention shifts away from the "substitute" territories and focuses on the anemone. Aggressive behavior and intraspecific encounters become more frequent around the anemone, and the largest fish generally begins driving the smaller ones away. Gradually these attacks increase in intensity and frequency, and the smaller fish usually must be removed from the aquarium to prevent injury or death. Eibl-Eibesfeldt (1960) similarly has reported that *Amphiprion percula* individuals remain friendly and school together until an anemone is introduced, whereupon fighting begins. Although detailed studies of this phenomenon using controlled experiments and models of different design have not been conducted, such studies should provide interesting comparative data on intraspecific and interspecific behavior, aggression, and territoriality, as well as on color and pattern discrimination in marine fishes.

c. Agonistic Behavior. With a few exceptions such as *Premnas* (Verwey, 1930), agonistic behavior appears to be causally related to territoriality; when the territory (i.e., the anemone) is removed, most intraspecific and interspecific aggression ceases. For example, two *Amphiprion xanthurus* individuals of approximately equal size will fight over an anemone until one is defeated. If the fish are of unequal size, the smaller one will usually give way to the larger after only a few encounters and will usually not be allowed to approach the anemone. In fact, the submissive fish must often be removed from the aquarium containing an anemone, since each sighting by the dominant fish results in renewed chases and fighting. Interspecific encounters are similar in that almost any fish approaching an anemone inhabited by an aggressive *Amphiprion* will often be attacked. This behavior is at variance with the idea that *Amphiprion* serve as a decoy to lure larger predatory fish into the anemone where they are stung and eaten. No such behavior has been observed (Mariscal, 1966*a*).

Attacks include biting, head-to-tail and tail-to-tail blows, and chasing. During chases the smaller fish often turns on its side and rapidly shakes or jerks its head from side to side three to five times. During the head shaking, audible sound in the form of spaced clicks can often be distinguished (Verwey, 1930; Koenig, 1960; Eibl-Eibesfeldt, 1960; Schneider, 1964; Mariscal, 1970*b*). Schneider (1964), the only worker so far to analyze electronically the sound spectra of various *Amphiprion* species, finds that he can break them down into three distinct types: threat sounds, battle sounds, and shaking sounds. These are in addition to several more generalized sounds produced during feeding and normal swimming. Such a sound repertory gives anemone fishes the largest "vocabulary" of any fishes so far investigated. It would be of interest to carry such studies further, using a wider range of species and making an analysis of each of the major *Amphiprion* behavior patterns. It would also be of interest to analyze such behavior in regard to the use of specific *Amhhiprion* color and banding patterns in agonistic and territorial displays.

d. Reproductive Behavior. A great deal of work on other organisms has been conducted in recent years which Aronson has reviewed up to 1957 and Reese to 1964 for marine fishes. However, courtship and reproductive behavior have not been analyzed in detail for *Amphiprion*, although there are scattered reports of breeding (mostly in aquaria) and juvenile development for some species (Verwey, 1930; Mitsch, 1941; Gohar, 1948; Garnaud, 1951; Davenport and Norris, 1958; Hackinger, 1959, 1962, 1967; Koenig, 1959, 1960; Oesmann, 1961; Springmann, 1963; Graefe, 1964; Graefe and Hackinger, 1967). Baerends (1957) discusses the ethological aspects of reproductive behavior in other fishes.

Either an adult pair of *Amphiprion* individuals or an adult pair plus juveniles of varying ages may occupy a single anemone, depending on the species of fish involved. The adult male–female pairs probably mate for life (Verwey, 1930). Some anemones and their fish are isolated from neighboring anemones by large areas of open water, making new colonization difficult. Where many anemones and fish occupy a single area, observations of easily recognizable pairs show that the pair bond remains intact. In all cases, the strong territorial behavior of most species would probably cause any new adult to be driven off by both residents, as I have observed in field and aquarium studies.

Regarding sexual dimorphism, in some species the adults are of unequal size, with the female being larger (e.g., *A. percula* and *Premnas biaculeatus*). In others, both adults are approximately the same size (e.g., *A. akallopisos A. ephippium*, and *A. xanthurus*). There is no general sexual dimorphism in color or pattern for most *Amphiprion* species. However, Valenti (1967) has reported that *A. percula* may be sexed by the width and configuration of the second white band, this being broader and more pointed anteriorly in the male.

Although a pair remains together throughout the year, reproduction may be restricted to certain seasons. Off Indonesia, this period appears to be associated with the shift in the monsoons (i.e., from April to June) (Verwey, 1930).

Contrary to de Crespigny's (1869) and Moser's (1931) idea that the eggs are laid on the anemone's disc, it has been found that they are commonly deposited on rocks near the base of the anemone (Hackinger, 1967). Just prior to egg laying, the male bites at and clears the substrate of algae and detritus. The eggs are laid at night, and both parents then begin to care for the brood, with the male doing most of the work. This primarily consists of cleaning the eggs by removing the detritus and settling organisms with the mouth and ventilating the water. In addition, Verwey reports, the male fish is able, by rubbing its body against the anemone's column, to cause the anemone to remain expanded and provide cover. Similar behavior both with stoichactiid anemones and *Calliactis* has already been mentioned. The eggs hatch at night, as Verwey (1930) and Hackinger (1967) mention, about 7 days after being laid. Initially the fry are planktonic, but they usually settle after 6–10 days to seek out an anemone. Hackinger (1967) discusses the initial cautious approaches of the young fishes to an anemone which are very reminiscent of the acclimation behavior of the adults. Gradually the juveniles penetrate the tentacles and finally take up residence there. This initial acclimation by young fishes has been little studied, and a detailed ethological analysis of the various stimuli involved in this behavior as compared with that of the adults, and in relation to various anemone species, would be of great interest.

e. Food and Feeding. Members of the genera *Amphiprion*, *Premnas*, and *Dascyllus*, observed both in the field and aquaria, appear to be largely omnivorous plankton feeders. This includes *Amphiprion akallopisos*, *A. percula*, *A. xanthurus*, *A. ephippium*, and *A. nigripes*. Any kind of net plankton or drifting organic material will be bitten at and seized as it passes the anemone. However, once such material touches the bottom, interest in it usually disappears. There is a certain degree of species specificity in this, since *A. ephippium* occasionally feeds on benthic material.

Amphiprion also feeds on waste material egested by the anemone (Sluiter, 1888; Seitz, 1926; Yonge, 1930; Verwey, 1930; Moser, 1931; Gohar, 1934, 1948; Koenig, 1960; Mariscal, 1966*b*, 1970*b*). I have observed on a number of occasions an *Amphiprion* actually poking its snout forcefully into the mouth of an anemone and removing mucus and various bits of organic debris which are then eaten. In addition, the normal movements of the symbiotic fish frequently result in sweeping the oral disc clean of organic and inorganic detritus.

A number of *Amphiprion* species also nibble (Fig. 2) and in some cases actually tear off and ingest large pieces of their host anemone's tentacles (Verwey, 1930; Eibl-Eibesfeldt, 1960; Mariscal, 1966*b*, 1970*b*). No apparent harm ensues from the ingestion of the nematocysts; in some cases these may pass through the digestive tract undischarged, while in others they appear to be digested along with the symbiotic algae (zooxanthellae) contained in the tentacles.

Perhaps one of the more interesting aspects of behavior displayed by some *Amphiprion* species is the taking of food to the anemone (Sluiter,

1888; Verwey, 1930; Gohar, 1934, 1948; Herre, 1936; Ladiges, 1939; Hackinger, 1959; Okuno and Aoki, 1959; Koenig, 1960; Graefe, 1964; Mariscal, 1966*a*,*b*, 1970*b*). I have observed this at the Green Island Aquarium on the Great Barrier Reef and hundreds of times since in both shipboard and laboratory aquaria (Fig. 3). Although the behavior is generally described as "feeding," a species such as *Amphiprion xanthurus* (with the behavior well-developed) will return to its anemone any drifting object too large to be immediately eaten and will push this material into the tentacles or deposit it directly on the oral disc (Fig. 3). No discrimination as to the organic or inorganic nature of such objects is made, and both types of material are returned to the anemone as soon as observed by the fish.

Gohar (1934, 1948) and Mariscal (1966*b*, 1970*b*) have observed an extreme example of this behavior: the feeding of living fishes to its anemone by an *Amphiprion*. Gohar reported that live sardines were attacked by the resident *Amphiprion bicinctus* and forced into the tentacles until subdued, whereupon they were eaten by the anemone.

I observed an *Amphiprion nigripes* engage in similar behavior. As an unprotected, living *Dascyllus trimaculatus* slowly swam toward the bottom

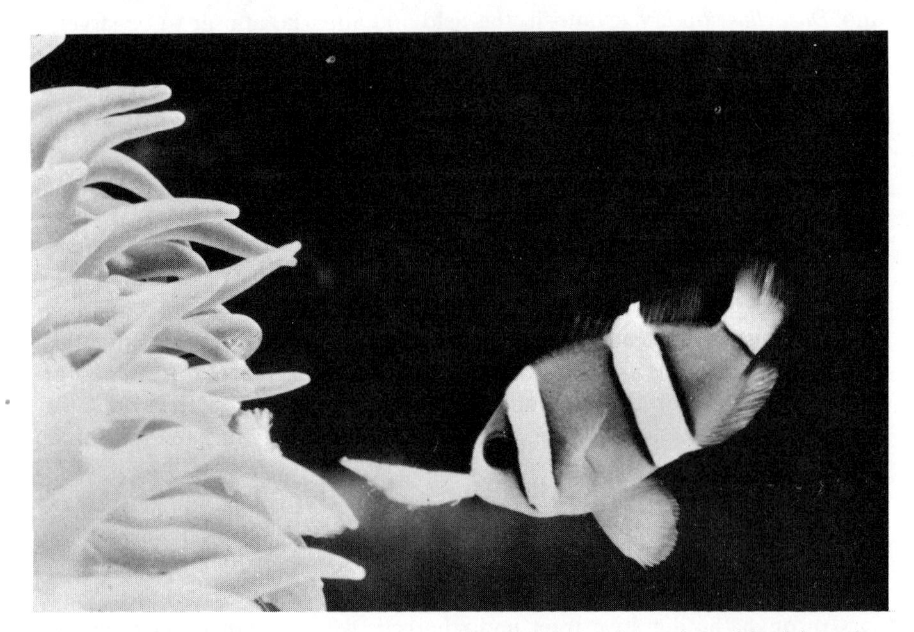

Fig. 3. Juvenile *Amphiprion xanthurus* (orange morph) conveying a piece of food from the opposite side of the aquarium, where it was introduced, to the tentacles of its sea anemone. Note the other pieces of food among the tentacles. This species of *Amphiprion* will return any type of drifting object (whether organic or inorganic) to the tentacles of its host anemone, provided the object is too large to be immediately ingested by the fish.

after being slightly stung by a *Radianthus ritteri* anemone, the resident *Amphiprion nigripes* dashed out and seized the *Dascyllus* just behind the head and immediately turned to thrust it deeply into the anemone's tentacles. The *Amphiprion* then turned away as the anemone's tentacles strongly adhered to and enveloped the fish. This was repeated a number of times with identical results. Dead fishes such as myctophids, gonostomatids, and atherinids were all fed to the same anemone by *A. nigripes.* It is likely that any non-*Amphiprion* species capable of being captured by a "feeder" *Amphiprion* would be returned to the host anemone. Normally, live fishes such as *D. trimaculatus* are agile enough to elude *Amphiprion* when attacked.

When its host anemone was removed, an *A. nigripes* persisted in attempting to feed an anemone which it was not allowed to enter due to the strong territorial defense of the resident *A. percula.* In one case, the *A. nigripes* spent 11 min attempting to place a piece of food among the tentacles of a *Radianthus ritteri* anemone before finally succeeding. It then immediately turned and left the anemone before being attacked by the resident fish.

In other experiments, pieces of food were dropped in the far end of an aquarium. When the resident *A. xanthurus* dashed out to seize it, its anemone was removed. As the fish returned with the food to find its anemone gone, it hesitated momentarily before carrying the food to the nearest corner, which was then "fed." This was the corner in which the fish had set up a territory prior to the introduction of the anemone. When the anemone was reintroduced, the fish again began feeding the anemone and neglected the corner.

f. Removing Inorganic Material from Around the Anemone. Verwey (1930) made the first and only report that an anemone fish (*Premnas biaculeatus*) enlarged a hole under its anemone by carrying away pieces of coral rubble. However, I have observed a similar phenomenon for *Amphiprion xanthurus* (Mariscal, 1966*b*, 1970*b*). Occasionally this fish would interrupt its normal "bathing" or swimming movements to halt and study something on the bottom near the edge of the anemone's oral disc. This was followed by the fish seizing a pebble and carrying it about 20–30 cm away from its anemone (*Stoichactis kenti*), where it was released. The fish then returned to the anemone. This was observed about a dozen different times for the same species. Thus, this species (*Amphiprion xanthurus*) not only returns objects (both organic and inorganic) to its anemone, but it also carries inorganic objects away and picks up organic waste from the disc. It would be interesting to study the various stimuli which trigger these apparently opposing behavioral patterns.

g. Preference of the Fishes for Specific Species of Anemones. Several authors have commented on the specificity and varying degrees of attachment of anemone fishes for their anemones in the field (Saville-Kent, 1893; Verwey, 1930; Eibl-Eibesfeldt, 1960; Mariscal, 1966*b*, 1970*b*). In some cases, it has been noted that fishes found with certain species of anemones in the field will accept other anemone species under aquarium conditions (Verwey, 1930; Coates, 1964; Blösch, 1965; Mariscal, 1965, 1966*b*, 1967, 1970*a,b*). However, this seems to hold true only for nonspecific *Amphiprion* species such as *A. xanthurus*, which the author has found in the field with seven different species of anemones from four genera and three families. In the aquarium, this was the only *Amphiprion* species tried, that readily accepted "foreign" anemones such as *Anthopleura* from California. *Amphiprion percula*, *A. frenatus*, *A. ephippium*, and *Premnas biaculeatus* would not acclimate to these same anemones after making an initial contact and being stung. Eibl-Eibesfeldt (1960), in field experiments in the Indian Ocean, and Davenport and Norris (1958), in aquarium experiments, found a similar situation. Eibl-eibesfeldt observed that *Amphiprion xanthurus* would voluntarily enter all three species of sea anemones in the study area, while *Amphiprion akallopisos* and *A. percula* would enter only the one in which they were normally found. When forced into the other two anemones, the latter fishes were badly stung.

Such specificity seems to occur repeatedly in certain geographic areas (see Table I). For example, I found *Amphiprion akallopisos* only with *Radianthus ritteri* off East Africa, the Seychelles Islands, and western Thailand. *Amphiprion percula* was not found in the western Indian Ocean, but is common in the western Pacific and eastern Indian Ocean. Off Thailand, *A. percula* was found only in *R. ritteri*, while in the Pacific it was common in *Stoichactis kenti* as well. In the Maldive Islands, *A. nigripes* was found only with *R. ritteri*. In all these areas, other anemone species and their characteristic *Amphiprion* were common and nearby, but I did not discover a random distribution of the adults among the various anemone species. In one case, I did find a juvenile *A. xanthurus* living with several *A. nigripes* individuals in a *R. ritteri* anemone in the Maldives. Although *R. ritteri* is common all across the Indian Ocean, for some reason *A. xanthurus* was not found with it (with this one exception), *Stoichactis giganteum* being the anemone of choice in this region. *Stoichactis kenti*, a common host anemone for *Amphiprion* in the Pacific, was not found in the Indian Ocean, where *Stoichactis giganteum* was common. Both *Amphiprion percula* and *Amphiprion xanthurus*, among others, were found with *S. kenti* in the Pacific. *Amphiprion ephippium* was found only with *Physobrachia ramsayi* in the eastern Indian Ocean, this anemone apparently being the normal host also for *Premnas biaculeatus* here and in the western Pacific, according to various reports (Verwey, 1930).

Based on personal observations as well as literature surveys, it seems clear that some very interesting problems in zoogeography and speciation of both anemone fishes and anemones throughout the Indo-Pacific region remain to be solved. More detailed studies and collections of anemones and their symbiotic fishes together are needed in the Philippine, Australian, and Indonesian archipelagoes.

h. Recognition of Anemones by Anemone Fishes. Visual stimuli appear to be primary in the recognition of anemones by anemone fishes (Verwey, 1930; Herre, 1936; Gohar, 1948; Davenport and Norris, 1958; Mariscal, 1966*b*, 1970*b*). Blösch (1965) conducted some interesting experiments which indicated that the color of the anemone could be remembered and used by the fish in relocating it. Blösch also verified Verwey's (1930) idea that the length of the tentacles relative to the shape and size of the anemone's oral disc is important in attracting an anemone fish. Choice experiments by the author showed no chemical recognition or preference by an *Amphiprion* for an anemone with which it had been living for an extended period prior to the experiment. Any preference between two anemones of the same size, shape, color, and species seemed to correlate with the relative degree of expansion of the oral disc at the time of the experiment. On the other hand, an *Amphiprion* will select its "own" species of anemone from among several other species in an aquarium and associate only with that species.

The importance of spatial orientation for various species of *Amphiprion* has been discussed earlier (e.g., Abel, 1960*c*). This undoubtedly plays a role in the orientation of an anemone fish toward an anemone or object with which it has had prior experience. It would be interesting to examine in detail the visual cues involved in such behavior, the length of time they are remembered, the time required to establish spatial discrimination, and the distance from which such cues can be distinguished.

i. Recognition between Fishes. Interspecific recognition between the various species of *Amphiprion* was quite striking (Mariscal, 1970*b*). For example, while *Amphiprion nigripes* would seize and return to its anemone dead fishes which had black or black-silver body markings, dead *Amphiprion percula* individuals with their three conspicuous white bars and bright orange body color were violently attacked and pushed away from the vicinity of the anemone's disc. It would be of interest to see what bar, pattern, and color combination causes an *Amphiprion* to reverse its behavior from that of feeding dead fishes to its symbiotic anemone to that of attacking and pushing them away from it.

Intraspecific recognition in the form of territorial behavior has also been commonly observed by the author. However, whether an *Amphiprion* is responding to species-specific stimuli or to a general *Amphiprion*-type pattern of vertical white bars against a contrasting body color remains to be demonstrated. As Thorpe (1963) has pointed out, there is no good experimental evidence for individual recognition in fishes. Even so, I have observed on several occasions what appeared to be clear cases of individual recognition among *Amphiprion xanthurus* specimens isolated from sea anemones in aquaria. Normally, such fish school together and, as discussed earlier, there is little or no agonistic behavior until an anemone is introduced. Occasionally, however, a single fish out of about five with the same color and markings would be singled out for attack by another fish of the group. In order to prevent serious injury, the attacked fish finally was removed to another aquarium. The attacks immediately stopped, with the remaining four fish schooling peacefully. However, upon reintroduction of the same fish following a separation of up to 1 week, the attacks again were initiated by the same aggressor. Individual recognition of one fish by the other seems clear in these cases, although the motivation for such behavior remains enigmatic.

It is curious that only a single species of anemone fish is usually found associating with a single anemone species in a geographically restricted area (Table 1). Since the newly hatched fry are planktonic during their first week of life, there is little possibility that the anemone they eventually select will be the same one that harbors their parents. In spite of this, for several species such as *A. percula*, *A. akallopisos*, and *A. nigripes*, there are often a number of small to large juveniles of all ages found in a single anemone along with the adults. The exact relationships here remain to be studied.

j. Species-Specific Behavior. Despite the basic similarity in their habits, there are distinct differences in behavior among the various species of anemone fishes. These include differences in territoriality, agonistic behavior, ranging behavior, flight behavior, feeding behavior, and sound production.

Regarding territoriality, some species of *Amphiprion* are commonly found in relatively large numbers in a single anemone. I have observed up to eight moderately sized *A. percula* or *A. nigripes* individuals in a single *Radianthus ritteri*. In addition to these easily visible fishes, there might have been several other tiny juveniles lost to sight among the tentacles. Verwey (1930) has commented on the apparently weaker territorial defense which he believes allows a large number of *A. percula* individuals to live together in a single anemone. However, in field as well as aquarium experiments, some *A. percula* individuals display territorial behavior similar to that of supposedly more aggressive species. On the other hand, I have observed seven or more *Amphiprion akallopisos* individuals inhabiting clusters of four to five *R. ritteri* anemones in the Seychelles Islands. In such anemones. subterritories are commonly set up so that one or two anemones in the cluster are particularly favored and defended against other fish in the cluster. Verwey (1930) has observed a similar situation for this same species of fish and anemone in Indonesia, as has Eibl-Eibesfeldt (1960) in the Nicobars.

Amphiprion akallopisos is a very territorial species; it frequently swam out and charged me as I approached its anemone. On one occasion, two *A. akallopisos* individuals delivered some surprisingly hard raps on the soles of my swim fins, which were close to their anemone, while several others attacked and bit at my hands and dip net as I was attempting to collect their host *R. ritteri*. In the Pacific, *Amphiprion perideraion*, a very close relative of the common Indian Ocean form *A. akallopisos*, was found with four or five individuals in a single anemone.

A pair of *A. xanthurus* individuals was often found with a host *Stoichactis*, but occasionally up to four or five relatively large fish might be associated with a single anemone. In the aquarium, *Amphiprion xanthurus* was very territorial and generally no more than a single fish and anemone could be established in a single aquarium. The author's hands or dip net were often attacked by *A. xanthurus*, the frequency and intensity of attacks being proportional to their distance from the anemone. *Amphiprion ephippium* also strongly defended its *Physobrachia* anemone against other *Amphiprion* individuals as well as fishes of other genera (e.g., *Dascyllus*) that approached too closely.

Differences in agonistic behavior, both intraspecific and interspecific, have been described by Verwey (1930), Eibl-Eibesfeldt (1960), and Schneider (1964). Eibl-Eibesfeldt (1960) reported that *A. percula* was more aggressive and always won battles with *A. akallopisos*. In attacks of *A. xanthurus* on *A. frenatus*, Schneider (1964) found that the latter was easily injured and always the loser in such encounters, while *A. percula*, being more agile, was able to elude the attacks of *A. xanthurus* but never able to drive off the latter from an anemone. Although very little work has been done on analyzing the actual behavioral acts involved in fighting from a comparative or descriptive point of view, Eibl-Eibesfeldt (1960) found that *A. akallopisos* lacked some of the typical behavioral patterns that *A. percula* displayed under similar conditions.

Regarding ranging behavior as a measure of the degree of attachment of an *Amphiprion* species for its anemone, distinct differences emerge. *Amphiprion percula* has been known for many years to form a very strong attachment to its anemone and rarely ranges more than a foot or so from it (Sluiter, 1888; Verwey, 1930; Ladiges, 1939; Eibl-Eibesfeldt, 1960; Mariscal, 1966*b*, 1970*b*). *Amphiprion akallopisos*, *A. perideraion*, *A. nigripes*, and *A. melanopus* also remain close to their anemones (Eibl-Eibesfeldt, 1960; Mariscal, 1966*b*, 1970*b*). Larger *Amphiprion* species are often more far-ranging. These include *A. ephippium*, *Premnas biaculeatus*, and especially *A. xanthurus*, which often swims out several meters from its anemone. Abel (1960*a*) has reported that *A. bicinctus* also exhibits a relatively loose association and may be found up to 4 m from its anemone.

Differences in flight behavior have been noted by various workers. In case of danger, *A. percula*, *A. akallopisos*, and *A. perideraion* all dive deeply into their respective anemone's tentacles and often may be collected in this way when the anemone is freed from the substrate (Eibl-Eibesfeldt, 1960; Mariscal, 1966*a*,*b*, 1970*b*). On the other hand, a larger, better swimmer such as *A. xanthurus* often flees from its anemone to hide in nearby coral when frightened, as Verwey (1930), Eibl-Eibesfeldt (1960), and I have all observed, and as Abel (1960*a*) has reported for *A. bicinctus* from the Red Sea. Verwey (1930) found, as I did, that *A. ephippium* also fled from its anemone on occasion but usually dove into other nearby anemones rather than holes in the coral, its strength of attachment being intermediate between that of *A. percula* and that of *A. xanthurus*.

There are also similar parallels in the "feeding" of the anemone by the fish. *Amphiprion percula* and *A. akallopisos* have not been observed to feed their anemones, whereas *A. xanthurus* does this each time food is presented. *Amphiprion ephippium* is again somewhere between these two extremes and feeds its anemone only about 50 percent of the times that food is presented to it. *Premnas biaculeatus* feeds its anemone slightly more than *A. ephippium* but much less than *A. xanthurus* (Verwey, 1930). I have found *A. nigripes* from the Maldive Islands to be as proficient at feeding the anemone as *A. xanthurus*. Apparently, *A. bicinctus* from the Red Sea also demonstrates this behavior (Gohar, 1934, 1948; Graefe, 1964).

Although it is not known how many species of *Amphiprion* produce sound, at least some can. Verwey (1930) reported that *A. xanthurus* and *A. akallopisos* definitely produced sound, while the other species he studied (*A. percula*, *A. ephippium*, and *Premnas biaculetus*) did not. However, Eibl-Eibesfeldt (1960) has reported that both *A. percula* and *A. akallopisos* produce some sounds when fighting, and I have heard this in *A. xanthurus*. Schneider (1964) reported that *A. xanthurus*, *A. polymnus*, *A. frenatus*, and *A. percula* all were capable of producing sound. He studied electronically the sound spectra of *A. xanthurus* and *A. polymnus* and found them to be nearly identical, as was the fighting behavior. It is possible that his *A. polymnus* was a juvenile form of *A. xanthurus*, since it is known that there is a great deal of variability within this species (Verwey, 1930; Mariscal, 1966*b*).

In addition to species-specific behavior, there is also much individual variation in behavior within a single species of *Amphiprion*. This has been observed in acclimation behavior, territorial behavior, agonistic behavior, and feeding behavior.

k. Nocturnal Behavior. At dusk, various species of *Amphiprion* and *Dascyllus* observed both in the shipboard and laboratory aquaria settle into their anemones, where they lean against a fold of the oral disc or against the upright tentacles until dawn. All locomotor activity appears to cease, and the distinctive coloration fades. The black and white pigmentation of a fish such as *Amphiprion xanthurus* tends to merge into a rather uniform grayish-brown color. Orange and reddish fish take on a paler color, while the black and white *Dascyllus* takes on a overall grayish color. This apparently is an adaptation of the fishes which allows them to blend more closely with their anemones, making them less conspicuous to a nocturnal reef predator. A predator presumably could more easily see and pluck the immobile anemone fish from among the amemone's tentacles without such "protective coloration."

2. Behavior of Fishes Isolated from Sea Anemones

a. General Condition. It has commonly been noted by the author as well as by importers of tropical marine fishes that disease is much more common, and deaths more frequent, among the various species of *Amphiprion* kept isolated from contact with sea amemones than among fishes kept with anemones. Fishelson (1965) has reported a similar observation. It is likely that a stress situation develops among isolated fishes due to lack of contact with a familiar anemone, which serves a number of important functions in the life of the fishes (Mariscal, 1966*a*). Although the effects of stress have been little investigated in lower vertebrates or invertebrates, a careful comparative study of *Amphiprion* behavioral patterns with and without sea anemones might provide valuable insights. The effects of stress in causing disease and, in some cases, death are well known for mammals and at least one insect (Ewing, 1967).

b. Acceptance of Substitute Anemones. In the absence of a sea anemone, captive *Amphiprion xanthurus* tended to adopt or set up territories in the vicinity of certain features of their aquarium. These areas were then defended, fed, settled in at night, and in general responded to as if they were sea anemones. Rieger (1962), Blösch (1965), and Mariscal (1966*b*, 1970*b*) have discussed this behavior.

One of the commonest behavioral traits of isolated *Amphiprion xanthurus* specimens was digging holes, usually in corners of the aquarium, which were "fed," were defended (although not nearly so vigorously as an anemone), served as refuges during the day, and were settled into at night. These are all patterns of behavior normally associated with the fish's symbiotic anemone. The holes were dug by the fish placing its snout against one end of the aquarium and then performing rapid swimming movements, the backwash of which threw the gravel away from under the fish. The same fish, when kept with an anemone, did not dig such holes. Blösch (1965) has reported similar observations for *A. xanthurus*.

When small bushy growths of green algae were placed in an aquarium containing isolated *Amphiprion xanthurus* specimens, some of the fish began settling into these as if they were anemones, as Rieger (1962) has mentioned for *A. percula*. The fish halted all motor activity and remained with mouth agape, leaning against the upright strands of algae for short periods. The fish would then swim out and circle to repeat the same behavior.

A third aspect of the behavior of isolated *Amphiprion xanthurus* specimens consisted of bathing in the bubbles produced by an air stone (Mariscal, 1966*b*, 1970*b*). The fish commonly backed into the air stream either completely, so that they were carried to the top of the aquarium, or partially, so that the air bubbles bounced off their flanks as the fish turned slightly from side to side (Fig. 4). Occasionally, the fish backed into the air stream and then attempted to swim against the current, emerging to repeat the process over again. Territoriality was displayed between two fish bathing in a single air stream, with the larger fish driving the smaller one aside. The impression obtained from detailed observation of this behavior was that the fish were receiving tactile stimulation from the impact of the air bubbles against their sides, perhaps similar to that received from brushing through an anemone's tentacles. When a *Stoichactis kenti* anemone was placed in the same aquarium, both fish soon began to acclimate to the anemone, paying no further attention to the air stream. Upon removing the anemone, the fish returned to bathing in the air stream. Other control *A. xanthurus* specimens of the same size living with an anemone in an adjacent aquarium never engaged in the air-bubble "bathing."

The most unexpected behavior of such isolated fish occurred when the air stone was shut off. The fish maintained their normal positions for several seconds and then began swimming around the air stone. About 30–60 sec after the bubbles ceased, the fish placed their mouths against either the air stone or the tubing and suddenly pushed these objects up to 6 cm across the aquarium. On resumption of the air flow, the fish ended these "attacks" and quickly assumed their former positions in the bubble stream. Subsequent disruption of the air supply caused renewed biting and pushing of the air stone. If the air supply remained shut off for more than 5 min, the biting ceased and the fish left the vicinity of the air stone. The possibility of some form of associative learning here is worthy of further study.

c. Territoriality. Territorial defense was exhibited by isolated fish, but to a lesser degree than when an anemone was present. Corners, holes, or other features were defended when a smaller fish passed too close, but there were no extended battles or even serious physical contact. Generally, the only response was a short dash or series of rushes at the second fish, these being quickly broken off when the intruder turned away. The spatial arrangement of the fish bathing in the stream of air bubbles also seemed due to territorial behavior of the larger fish.

Fig. 4. Pair of juvenile *Amphiprion xanthurus* specimens "bathing" in the air bubble stream. This only occurred among *Amphiprion* specimens which had been isolated from any contact with sea anemones for several days. When an anemone was introduced, the fish soon left the air stream to bathe among the tentacles of the anemone and the air stream was ignored. The larger fish (to left) was dominant and habitually frequented the left side of the air stream, which it defended against the other fish.

d. Agonistic Behavior. Except on rare occasions, as discussed earlier, there was little agonistic behavior exhibited by isolated fish except during feeding and the slight territorial defense. A number of *Amphiprion* (six to 12) specimens of the same or different species could be kept in a single aquarium with little difficulty. During feeding, however, the smaller fishes were often harrassed by the larger fishes and driven away.

e. Feeding Behavior. Isolated *Amphiprion xanthurus* specimens returned organic as well as inorganic materials to the corner of the aquarium in which they had dug a hole. They made repeated attempts to push the food into the corner and then released it as they did with an anemone. Of course, in the absence of nematocysts the food began to sink to the bottom, whereupon the fish would grasp it again and repeat the same behavior. The aquarium heater, outside filter siphon tube, and inside filter were all "fed" in similar fashion. Finally, after a number of such futile attempts spanning several minutes, the food was allowed to fall to the bottom. However, if the same piece was picked up and redropped into the aquarium, feeding behavior was elicited again and continued as long as the fish saw the food before it reached bottom. Three to four dozen successive feedings could be elicited before the experimenter tired; the fish seemed unaffected. Even the air-bubble stream was "fed" by the fish which were associated with it. The food was carried across the aquarium and released just inside the stream of bubbles. As the current caught the food, the fish turned away or began bathing in the stream again. Occasionally, the same piece of food would be carried up and out of the bubble stream and into the line of vision of a fish, whereupon it again seized the food and repeated the feeding behavior. In nearly all cases, only pieces of food too large to swallow easily were returned; small pieces were usually ingested immediately.

f. Rocking Behavior. This consists of a rapid elevation and depression of the anterior portion of the body while the fish remains in place, and has been observed for most species of *Amphiprion*. It has been called "up-and-down swimming," "bobbing," "bouncing," and "seesawing" behavior, among other things, and appears to be more pronounced among isolated fishes, although it occurs in those fishes kept with anemones as well (Eibl-Eibesfeldt, 1960; Mariscal 1970*b*). Although its function is still unclear, it is possible that it may enable a fish to range on and estimate the distance of an approaching object before finding shelter. Various insects and birds have been observed by the author to engage in this behavior under similar circumstances. Another possibility is that this behavior is analogous to the "push-up" territorial displays commonly seen among iguanid lizards, although *Amphiprion* individuals without anemones or any well-defined territory also engage in this behavior. It may also serve a visual discrimination function in lizards, although this possibility has not been investigated to my knowledge.

g. Exploratory and Play Behavior. Although play and exploratory behavior have not been convincingly demonstrated for fishes, there are some suggestive possibilities (Thorpe, 1963). As Berlyne (1966) has clearly pointed out, many higher animals tend to expend considerable time and energy seeking stimuli which appear to have no primary ecological significance for their survival. Such behavior has generally been considered under the headings of curiosity, play, or exploratory behavior.

In considering the behavior of anemone fishes, and in particular isolated anemone fishes, I have observed several types of behavior which very tentatively might be included under these headings. For example, the rocking behavior of anemone fishes, both with and without anemones and both in the laboratory and the field, appears more pronounced when a foreign object enters the fish's field of view. This behavior appears to be correlated with an increasing degree of "uncertainty" on the part of the fish whether to maintain its position or to flee to the shelter of the tentacles. *Amphiprion* specimens kept without anemones appear, in some cases, to carry out this behavior continually as they move about the aquarium, and in general it seems more pronounced than when the same fishes are kept with anemones. Such behavior might fit into the general category of specific exploratory behavior, as discussed by Berlyne (1966).

Second, and more interesting, is the behavior of isolated fishes in establishing what I have called "substitute anemones." As mentioned earlier, this consists of hole-digging, settling into tufts of algae, and orienting to a stream of air bubbles. This behavior might be included under what Berlyne (1966) calls "diversive exploratory behavior." Perhaps the isolated anemone fishes are seeking "stimulation that taxes the nervous system to the right extent," the right extent in this case being the physical and psychological stimuli provided by the host anemone with regard to protection, degree of concealment, and perhaps tactile stimulation. The additional facts that the isolated fishes return food to their substitute anemones, settle in and against them at night (when the fishes become completely inactive), and defend them, all indicate that the fishes are responding to these features of their environment as if they were, in fact, living sea anemones. Obviously, any discussion of possible play or exploratory behavior by anemone fishes is highly speculative at the present time, but there are enough interesting possibilities to suggest further study.

IV. DISCUSSION AND CONCLUSIONS

One interesting finding of the present study is the apparent specificity of certain *Amphiprion* species for certain anemone species (Table 1). It would not have been surprising to find the various species of *Amphiprion* randomly distributed among the various species of large anemones present in a typical Indo-Pacific reef area. This was not the case. In any given area, generally only a single species of *Amphiprion* was found inhabiting a single anemone species. This was true even if other species of anemones were present. In such cases, these anemones also contained their own species of fish (Table 1). In another part of the world, however, the same species of fish might be found with a different species of anemone. For example, in the Indian Ocean I found *A. percula* only with *R. ritteri*, whereas in the Pacific I found this fish more commonly with *Stoichactis kenti*.

Amphiprion xanthurus was one of the most nonspecific species in its choice of anemone partners, but it was notably absent from one of the commonest Indo-Pacific anemones, *Radianthus ritteri*. In only one case was a single juvenile *A. xanthurus* found sharing this anemone with *A. nigripes* in the Maldives. The dozens of other *A. xanthurus* there were all found exclusively with *Stoichactis giganteum*, an anemone with which *A. nigripes* never associated. *Amphiprion nigripes* was found exclusively with *R. ritteri* in the Maldives. *Amphiprion ephippium*, on the other hand, was never found with either *Stoichactis* or *Radianthus* in the islands off the coast of western Thailand (although both genera were common), but instead was found exclusively with *Physobrachia ramsayi* (Table 1). Here *A. xanthurus* again was found exclusively with *Stoichactis giganteum*, and both *A. percula* and *A. akallopisos* were found with *R. ritteri* (but not together). Obviously, in all these cases, more extensive field observations might reveal a different pattern but would probably not affect the conclusion that, for any single locality, certain species of anemone fishes are invariably found associating with certain species of anemones. The various ecological, physiological, and behavioral factors which underlie this specificity are at present unknown.

Another interesting generality which emerged during the present study is that many of the giant Indo-Pacific anemones and their symbiotic fishes appeared to be more common in dead or less actively growing reef areas. The various ecological parameters which might influence this distribution, such as competition with the corals for space, food, and sunlight remain to be investigated.

One conclusion of the present survey is that the various species of *Amphiprion*, and probably *Premnas*, appear to form a rather close bond with their anemones, generally leaving them only when they fail to give the fishes reasonable concealment and protection. It is thus likely that the various species of *Amphiprion* and *Premnas* are obligatory symbionts in that they spend nearly their entire life span (with the exception of the short post-hatching planktonic stage) in or around sea anemones.

Another group of fishes which I have said little about, but which are deserving of further study, are those which might be called "facultative anemone symbionts." These include pomacentrids such as *Dascyllus trimaculatus*, found throughout the Indo-Pacific (Abel, 1963; Mariscal, 1970*b*), and the very closely related *Dascyllus albisella* from Hawaii (Stevenson, 1963, Mariscal, 1966*b*). Both species may associate with sea anemones as juveniles but are free-living as adults. Even as juveniles, however, both species of *Dascyllus* are not restricted to sea anemones but may be commonly found around branching stands of coral, into which they flee much as they do into the tentacles of sea anemones. Stevenson (1963) has found that *D. albisella* acclimates to sea anemones much like *Amphiprion,* and I have been able to confirm this as well as observe the same phenomenon for *D. trimaculatus.*

Another interesting problem which has not been studied in detail is the initial orientation and acclimation of newly hatched *Amphiprion* fry to sea anemones. Are there specific visual or chemical stimuli which attract certain species of *Amphiprion* to certain species of sea anemones, as appears to be the case for some species, or is it strictly a matter of chance? What factors underlie the apparent specificity of certain *Amphiprion* species for certain anemone species? How does the territoriality of *Amphiprion* species influence this initial acclimation and ultimate distribution of the fishes? What population of fish or fishes will a single anemone or a single reef area support and what are the underlying ecological factors in such distributions? Why is one species of *Amphiprion* apparently restricted to a single area (e.g., *Amphiprion nigripes* in the Maldive Islands), while others are found throughout the Indo-Pacific? There are many interesting questions which one can pose in studying such a symbiosis. Some can be solved only by extensive field observations, others only by laboratory study.

Although the present study tends to provide more questions than answers, the study of symbiotic associations may furnish us with an especially effective tool for the study of behavior *per se*, as Davenport (1966) points out. The importance of the respective roles of photoreception, mechanoreception, and chemoreception in the basic behavioral processes of reproduction, locomotion, orientation, feeding, fighting, and so on might be most suitably studied using symbiotic organisms. This is because a symbiotic association provides us with a "natural" arena for the study of behavioral interactions among organisms, both intraspecifically and interspecifically, as well as interphyletically. I have tried to show how some of these problems can be approached using a vertebrate–invertebrate association, that of pomacentrid fishes and sea anemones. I believe that enough background information is now available so that exhaustive analyses of the behavioral interactions of such symbioses will be not only possible, but rewarding.

REFERENCES

Abel, E. F., 1960*a*, Zur Kenntnis des Verhaltens und der Ökologie von Fischen an Korallenriffen bei Ghardaqa (Rotes Meer), *Z. Morph. Ökol. Tiere* **49:** 430.

Abel, E. F., 1960*b*, Liaison facultative d'un poisson (*Gobius bucchichii* Steindachner) et d'une anémone (*Anemonia sulcata* Penn.) en Méditerranée, *Vie et Milieu* **11:** 517.

Abel, E. F., 1960*c* Ein Beispiel für die räumliche Orientierung von *Amphiprion percula* (Lac.), *Die Pyramide* **3:** 78.

Abel, E. F., 1963, Anemonen und Anemonen-fische, *Neptun* **3(9):** 246.

Allee, W. C., Park, O., Emerson, A E., Park, T., and Schmidt, K. P., 1949, "Principles of Animal Ecology," W. B. Saunders Co., Philadelphia.

Araga, C., 1964, An anemone fish and chaetodont fish, new to the Japanese fish fauna, *Publ. Seto Mar. Biol. Lab.* **12(1):** 113.

Aronson, L. R., 1957, Reproductive and parental behavior, *in* "The Physiology of Fishes" (M. E. Brown, ed.) Vol. 2, pp. 271–304, Academic Press, New York.

Baerends, G. P., 1957, The ethological analysis of fish behavior, *in* "The Physiology of Fishes" (M. E. Brown, ed.) Vol. 2, pp. 229–269, Academic Press, New York.

Berlyne, D. E., 1966, Curiosity and exploration, *Science* **153:** 25.

Blösch, M., 1961, Was ist die Grundlage der Korallenfisch-symbiose: Schutzstoff oder Schutzverhalten? *Naturwissenschaften* **48(9):** 387.

Blösch, M., 1965, Untersuchungen über das Zusammenleben von Korallenfischen (*Amphiprion*) mit Seeanemonen, Inaugural-Dissertation, Eberhard-Karls-Universität zu Tübingen.

Bowman, T. E., and Mariscal, R. N. 1968, *Renocila heterozota*, a new cymothoid isopod, with notes on its host, the anemone fish, *Amphiprion akallopisos*, in the Seychelles, *Crustaceana* **14(1):** 97.

Buhk, F., 1939, Lebensgemeinschaft zwischen Riesenseerose und Korallenfischen, *Wchschr. Aquar. Terrarienkunde* **46:** 672.

Caspers, H., 1939, Histologische Untersuchungen über die Symbiose zwischen Aktinien und Korallenfischen, *Zool. Anz.* **126:** 245.

Coates, C. W., 1964, Safe hiding places moved while you wait, *Animal Kingdom* **67:** 77.

Collingwood, C., 1868*a*, Note on the existence of gigantic sea-anemones in the China Sea, containing within them quasi-parasitic fish, *Ann. Mag. Nat. Hist. Ser.* **4(1):** 31.

Collingwood, C., 1868*b*, "Rambles of a Naturalist on the Shores and Waters of the China Sea," John Murray, London.

Davenport, D., 1955, Specificity and behavior in symbioses, *Quart. Rev. Biol.* **30:** 29.

Davenport, D., 1966, The experimental analysis of behavior in symbioses, *in* "Symbiosis" (S. M. Henry, ed.) Vol. 1, pp. 381–429, Academic Press, New York.

Davenport, D., and Norris, K. S., 1958, Observations on the symbiosis of the sea anemone *Stoichactis* and the pomacentrid fish *Amphirpion percula*, *Biol. Bull.* **115:** 397.

de Bary, A. 1879, "Die Erscheinung der Symbiose," Trubner, Strassburg.

de Crespigny, C. C., 1869, Notes on the friendship existing between the malacopterygian fish *Premnas biaculeatus* and the *Actinia crassicornis*, *Proc. Zool. Soc. London* **1869:** 248.

Eibl-Eibesfeldt, I., 1960, Beobachtungen und Versuche an Anemonenfischen (*Amphiprion*) der Malediven und der Nicobaren, *Ztschr. Tierpsychol.* **17:** 1.

Eibl-Eibesfeldt, I., 1965, "Land of a Thousand Atolls," MacGibbon and Kee, London.

Eibl-Eibesfeldt, I., 1967, Formen der Symbiose, *Naturwiss und Med.* **16:** 14.

Ewing, L. S., 1967, Fighting and death from stress in a cockroach, *Science* **155:** 1035.

Fishelson, L., 1965, Observations on the Red Sea anemones and their symbiotic fish *Amphiprion bicinctus*, *Bull. Sea Fish Res. Stat. Haifa* **39:** 1.

Fricke, H. W., 1966, Attrappenversuche mit einigen plakatfarbigen Korallenfischen im Roten Meer, *Ztschr. Tierpsychol.* **23(1):** 4.

Garnaud, J., 1951, Nouvelles données sur l'ethologie d'un pomacentridé: *Amphiprion percula* Lacépède, *Bull. Inst. Océanog. Monaco* **998:** 1.

Gohar, H. A. F., 1934, Partnership between fish and anemone, *Nature* **134:** 291.

Gohar, H. A. F., 1948, Commensalism between fish and anemone (with a description of the eggs of *Amphiprion bicinctus* Rüppell), *Fouad I. Univ. Publ. Mar. Biol. Sta. Ghardaqa* (*Red Sea*) **6:** 35.

Graefe, G., 1963, Die Anemonen-Fisch-Symbiose und ihre Grundlage—nach Freilanduntersuchungen bei Eilat/Rotes Meer, *Naturwissenschaften* **50(11):** 410.

Graefe, G., 1964, Zur Anemonen-Fisch-Symbiose, nach Freilanduntersuchungen bei Eilat/Rotes Meer, *Ztschr. Tierpsychol.* **21(4):** 468.

Graefe, G., and Hackinger, A., 1967, Die Jugendentwicklung des Anemonenfisches *Amphiprion bicinctus*, *Natur und Mus.* **97(5):** 170.

Gudger, E. W., 1941, Coelenterates as enemies of fishes. IV. Sea anemones and corals as fish eaters, New Eng. Naturalist **10:** 1.
Hackinger, A., 1959, Freilandbeobachtungen an Aktinien und Korallenfischen, *Mitteil. Biol. Stat. Wilhelminenberg Wien* **2:** 72.
Hackinger, A., 1962, Die Entwicklung der weissen Streifenzeichnung bei *Amphiprion percula, Die Pyramide* **10(3):** 126.
Hackinger, A., 1967, Anemonenfische—im Aquarium gezüchtet, *Aquarien Mag.* **4:** 136.
Henry, S. M., (ed.) 1966, Associations of microorganisms, plants and marine organisms, *in* "Symbiosis", Vol. 1, Academic Press, New York.
Herre, A. W., 1936, Some habits of *Amphiprion* in relation to sea anemones, *Copeia* **1936(3):** 167.
Horst, R., 1903, On a case of commensalism of a fish (*Amphiprion intermedius*) and a large sea anemone (*Discosoma* sp.) *Notes Leyden Ms.* **23:** 180.
Koenig, O., 1959, Die Haltung von Riesenaktinien und Korallenfischen, *Mitteil. Biol. Stat. Wilhelminenberg Wien* **2:** 60.
Koenig, O., 1960, Verhaltensuntersuchungen an Anemonenfischen, *Die Pyramide* **8(2):** 52.
Ladiges, W., 1939, Das Rätsel der Symbiose zwischen den Riesenseerosen der Gattung *Stoichactis* und den Fischen der Gattungen *Premnas* und *Amphiprion*, *Wchschr. Aquar. Terrarienkunde* **46:** 669.
Mariscal, R. N., 1965, Observations on acclimation behavior in the symbiosis of anemone fish and sea anemones, *Am. Zoologist* **5(4):** 694.
Mariscal, R. N., 1966*a*, The symbiosis between tropical sea anemones and fishes: A review, *in* "The Galápagos" (R. I. Bowman, ed.) pp. 157–171, University of California Press, Berkeley.
Mariscal, R. N., 1966*b*, A field and experimental study of the symbiotic association of fishes and sea anemones, Ph. D. dissertation, University of California, Berkeley, University Microfilms, Ann Arbor, Mich.
Mariscal, R. N., 1967, A field and experimental study of the symbiotic association of fishes and sea anemones, *Dissertation Abs.* **28(1):** 388–B.
Mariscal, R. N., 1969, The protection of the anemone fish, *Amphiprion xanthurus*, from the sea anemone, *Stoichactis kenti*, *Experientia* **25:** 1114.
Mariscal, R. N., 1970*a*, An experimental analysis of the protection of *Amphiprion xanthurus* Cuvier & Valenciennes and some other anemone fishes from sea anemones, *J. Exp. Mar. Biol. Ecol.* **4:** 134.
Mariscal, R. N., 1970*b*, A field and laboratory study of the symbiotic behavior of fishes and sea anemones from the tropical Indo-Pacific, *Univ. Calif. Publ. Zool* **91:** 1.
Mariscal, R. N., 1970*c*, The nature of the symbiosis between Indo-Pacific anemone fishes and sea anemones, *Mar. Biol.* **6(1):** 58.
Mariscal, R. N., 1971, Experimental studies on the protection of anemone fishes from sea anemones, *in* "The Biology of Symbiosis" (T. C. Cheng, ed.) University Park Press, Baltimore.
Mitsch, H., 1941, Breeding of marine clown fishes, *The Aquarium* (*Philadelphia*) **10:** 48.
Moser, J., 1931, Beobachtungen über die Symbiose von *Amphirpion percula* mit Aktinien, *Sitzungsber. Ges. Naturforsch. Freunde Berlin* **1931(2):** 160.
Noble, E. R. and Noble, G. A., 1964, "Parisitology. The Biology of Animal Parasites." 2nd Ed., Lea and Febiger, Philadelphia.
Oesman, H., 1961, Erlebnisse mit Anemonenfischen, *Aquar. Terr. Ztschr.* **14(1):** 49.
Okuno, R., 1963, Observations and discussions on the social behaviors of marine fishes, *Publ. Seto Mar. Biol. Lab.* **11(2):** 281.
Okuno, R. and Aoki, T., 1959, Some observations on the symbiosis between the pomacentrid fish and the sea anemone, *J. Japan. Assoc. Zool. Gardens Aquar.* **1(1):** 8.
Reese, E. S., 1964, Ethology and marine zoology, *in* "Oceanography and Marine Biology Annual Review" (H. Barnes, ed.) pp. 455–488, George Allen and Unwin Ltd., London.
Rieger, W., 1962, Beobachtung an *A. percula. Aquar. Terr. Ztschr.* **15:** 221.
Ross, D. M. and Sutton, L., 1968, Detachment of sea anemones by commensal hermit crabs and by mechanical and electrical stimuli, *Nature* **217:** 380.

Saville-Kent, W. S., 1893, "The Great Barrier Reef of Australia," W. H. Allen & Co., Ltd., London.

Saville-Kent, W. S., 1897, "The Naturalist in Australia," Chapman and Hall, Ltd., London.

Schlichter, D., 1967, Zur Klärung der "Anemonen-Fisch-Symbiose," *Naturwissenschaften* **54:** 569.

Schlichter, D., 1968, Das Zusammenleben von Riffanemonen und Anemonenfischen, *Ztschr. Tierpsychol.* **25:** 933.

Schneider, H., 1964, Bioakustische Untersuchungen an Anemonenfische der Gattung *Amphiprion* (Pisces), *Z. Morphol. Ökol. Tiere* **53:** 453.

Schultz, L. P., 1953, Review of the Indo-Pacific anemone fishes, genus *Amphiprion*, with descriptions of two new species, *Proc. U.S. Natl. Ms.* **103:** 187.

Schultz, L. P., 1966, A new anemone fish, *Amphiprion calliops*, from the Indo-Pacific Oceans, *Ichthyologica* **37(2):** 71.

Schultz, L. P., Chapman, W. M., Lachner, E. A., and Woods, L. P., 1960, Fishes of the Marshall and Marianas Islands, *Bull. U.S. Natl. Ms.* **202:** 1.

Seitz, C., 1926, Lebensgemeinschaft zwischen Korallenfischen und Seerosen, *Bl. Aquar.* **37:** 413.

Sluiter, C. P., 1888, Ein merkwürdiger Fall von Mutualismus, *Zool. Anz.* **11:** 240.

Smith, D., Muscatine, L., and Lewis, D., 1969, Carbohydrate movement from autotrophs to heterotrophs in parasitic and mutualistic symbiosis, *Biol. Rev.* **44:** 17.

Springmann, W., 1963, Misslungene Zucht von *Amphiprion percula*, *Aquar. Terr. Ztschr.* **16(6):** 174.

Stephenson, T. A., 1946, Coral reefs, *Endeavour* **5:** 96.

Stevenson, R. A., 1963, Behavior of the pomacentrid reef fish *Dascyllus albisella* Gill in relation to the anemone *Marcanthia* (*sic*) *cookei*, *Copeia* **1963(4):** 612.

Thorpe, W. H., 1963, "Learning and Instinct in Animals," Methuen, London.

Valenti, R. J., 1967, Clown fish sexing: *Amphiprion percula*, *Salt Water Aquar.* **3(3):** 61.

Verwey, J., 1930, Coral reef studies. I. The symbiosis between damselfishes and sea anemones in Batavia Bay, *Treubia* **12(3–4):** 305.

Whitley, G. P., 1932, Fishes, *Brit. Mus.* (*Nat. Hist.*) *Great Barrier Reef Exp.* (**1928–29**) *Sci. Rep.* **9:** 267.

Yonge, C. M., 1930, "A Year on the Great Barrier Reef," Putnam, London.

Chapter 10

ACOUSTIC DISCRIMINATION BY THE TOADFISH WITH COMMENTS ON SIGNAL SYSTEMS

Howard E. Winn

Graduate School of Oceanography
University of Rhode Island
Kingston, Rhode Island

I. INTRODUCTION

The oyster toadfish, *Opsanus tau*, produces two basic sounds which have a variety of properties. The grunt is made by both sexes, but more frequently by males during aggressive encounters in the reproductive season. The boatwhistle call is produced by males in long spontaneous sequences on their nests. They do so most often at the beginning of the mating season. Individual grunts are sometimes spontaneously interspersed between boatwhistles. It is thought by Fish (this volume) and Winn (1967) that some of these are incomplete boatwhistles. Males increase their boatwhistle calling when boatwhistles are played back at certain rates (Winn, 1967). It was this response that was utilized by Fish (this volume) to elucidate the properties of pattern, interval, cycle time, and antiphony in *O. tau*. Here I will describe the physical parameters of the boatwhistle call, such as frequency, amplitude, rate, and duration, that result in vocal facilitation. These will be generally discussed in relation to the development of acoustic signal systems in fishes.

This research was supported by U.S.P.H.S. grant NB-03241 and ONR contract No. N00014–68–A–0215–0003 URI Oceanography.

One set of experiments on the functional significance of the boatwhistle call is also presented with a discussion of message content.

II. MATERIALS AND METHODS

Experiments were performed in 1966, 1967, and 1968 to establish what physical parameters of a sound influenced rates of boatwhistling. Tiles were placed adjacent to pilings along the pier at the Chesapeake Biological Laboratory, Solomons Island, Md., and the same playback equipment was utilized as described previously (Winn, 1967; Fish, this volume). A series of sound playbacks was presented to single fish to test their effectiveness in altering the fish's calling rate. Each series included several experimental playbacks and two controls of normal boatwhistle sounds at a stimulatory rate. One control playback was randomly ordered between the experimental tests; the other was presented at the end of each replicate of a series. The latter test determined if the test fish could still be stimulated to increase its calling rate after exposure to each replicate set of a series. The controls and some other experiments were organized in a standard way with a duration of 300 msec and a period of 2300 msec. All of the tests have been reordered in Table I, but each series was as follows (each experiment of a series played back in random order to *N* number of fish):

Series 2A—(Experiment No. in Table I) 12, 13, 1, 2, 3, 4, 5
2B—10, 11, 38, 39
1A—6, 7, 8, 16, 20, 21, 29, 30, 31
1B—9, 17, 18, 19, 32
4 —33, 34, 49, 50, 51, 52, 53, 54
5 —14, 15, 22, 23, 24, 26, 27, 35, 36, 37
6 —25, 28, 40, 41, 42, 43, 44, 45, 46, 47, 48

The artificial calls were made with an audio oscillator, a Brason-Stadler electronic switch (829E), and a tape recorder. They were visually monitored on an oscilloscope, and sonagraphic analyses were made of the signals recorded underwater in the actual experiments.

Sounds were recorded from an animal in each test for a 3 min preplayback period, a 3 min period when sounds were played out underwater, and a final 3 min postplayback period. The results of postplaybacks were given previously (Winn, 1967), and because no new information was obtained the changes in postplaybacks are not presented. In order for a test to be considered, the fish was required to call at least three times in the preplayback period. This criterion aided in standardizing motivational levels. Also, it was possible for a fish to call at his maximum rate during preplayback

Table I. Playback of Artificial Calls to Toadfish (See *Materials and Methods* for Further Detail)

	Playback call					Fish responses							
Experiment No.	*N*	Frequency (Hz)	Duration (msec)	Period (msec)	Interval (msec)	Preplay average	Playback average	Average difference	Number increased	Number decreased	No change	Significant sign test	Footnote
1	15	200	300	2000	1700	25	36	+11	12	3	0	0.05	
2	15	300	300	2000	1700	26	30	+4	10	5	0	—(0.05)	*a*
3	15	400	300	2000	1700	22	28	+6	12	2	1	0.01	
4	14	600	300	2000	1700	25	26	+1	8	6	0	—	
5	15	800	300	2000	1700	23	20	−3	8	7	0	—	
6	14	180/360/540	300	2300	2000	29	34	+5	12	2	0	0.01	*b*
7	16	180/360	300	2300	2000	21	30	+9	13	3	0	0.01	*b*
8	16	180/540	300	2300	2000	32	31	−1	10	6	0	—(0.05)	*b*
9	14	180/360/540 720/900	300	2300	2000	22	26	+4	9	5	0	—(0.01)	*b*
10	13	200	300	1900	1600	27	43	+16	12	1	0	0.01	*c*
11	14	400	300	2000	1700	27	36	+9	11	3	0	0.05	*c*
12	16	50(100)	300	2000	1700	19	12	−7	3	12	0	—	
13	15	100	300	2000	1700	21	21	0	7	6	2	—	
14	16	90	300	2300	2000	20	23	+3	8	7	1	—	
15	16	90	300	2300	2000	22	17	−5	6	10	0	—	*d*
16	15	90/270	300	2300	2000	23	17	−6	6	9	0	—	
17	15	90/180/270 360/450	300	2300	2000	21	27	+6	9	4	2	—	*b*
18	15	90/180/270	300	2300	2000	23	18	−5	6	9	0	—	*b,e*
19	15	90/180/270	300	2300	2000	28	13	−15	2	13	0	—	*b,e*
20	14	90/180/270	300	2300	2000	26	19	−7	3	9	2	—	*b*
21	16	90/180	300	2300	2000	27	23	−4	5	9	2	—	*b*
22	16	90/180	300	2300	2000	22	28	+6	14	1	1	0.01	*b*
23	16	90/180	300	2300	2000	25	30	+5	13	3	0	0.01	*b*
24	16	90/180	300	2300	2000	19	22	+3	9	7	0	—	*b,d*
25	15	growl	300	2300	2000	23	28	+5	10	5	0	—	*f*
26	16	200	75	2300	2225	22	25	+3	9	7	0	—	
27	16	200	75	2300	2225	22	14	−8	3	13	0	—	*d*

Table I (Cont'd)

Experiment No.	Playback call					Fish responses							
	N	Frequency (Hz)	Duration (msec)	Period (msec)	Interval (msec)	Preplay average	Playback average	Average difference	Number increased	Number decreased	No change	Significant sign test	Footnote
28	15	Boatwhistle	75	2300	2225	20	10	−10	1	12	2	—	*g*
29	16	180/360/540	150	1150	1000	27	21	−6	6	10	0	—	*b*
30	16	180/360	150	1150	1000	24	22	−2	9	7	0	—	*b*
31	15	180/540	150	1150	1000	24	20	−4	5	10	0	—	*b*
32	16	180/360/540 720/900	150	1150	1000	29	29	0	11	5	0	—	*b*
33	15	200	300	1200	900	27	31	+4	11	4	0	0.05	
34	15	400	110	1200	1090	29	32	+3	8	5	2	—	
35	16	200	75	1150	1075	22	32	+10	16	0	0	0.01	
36	16	90	Continuous tone			19	1	−18	0	16	0	—	
37	16	90	Continuous tone			13	1	−12	0	16	0	—	
38	14	200	Continuous tone			27	4	−23	0	14	0	—	
39	14	400	Continuous tone			25	8	−17	1	13	0	—	
40	15	200 + tone	300	2500	2200	22	5	−17	1	14	0	—	*h*
41	15	200 + tone	300	2500	2200	20	6	−14	1	14	0	—	*d,h*
42	15	200 + tone	300	5000	4700	21	4	−17	0	15	0	—	*h*
43	15	200	300	2700	2400	21	31	+10	13	2	0	0.01	*i*
44	15	200	300	2700	2400	22	31	+ 9	13	2	0	0.01	*i*
45	13	200	300	5000	4700	21	26	+ 5	10	3	0	0.05	*i*
46	15	200	300	2500	2200	20	34	+14	14	0	1	0.01	*i*
47	15	200	300	2500	2200	19	32	+13	15	0	0	0.01	*i*
48	15	200	300	5000	4700	18	26	+8	14	1	0	0.01	*i*
49	15	200	300	1100	900	21	38	+17	15	0	0	0.01	
50	15	200	500	1200	700	29	43	+14	14	1	0	0.01	
51	15	200	1000	1200	200	29	29	0	7	8	0	—	
52	15	200	150	1200	1050	27	33	+6	11	3	1	0.05	
53	15	200	500	1500	1000	23	39	+16	14	0	1	0.01	
54	16	200	1000	1900	1100	24	36	+12	12	3	1	0.05	

Series													
2A	14	Control	300	2300	2000	19	33	+14	13	1	0	0.01	j
2A	14	Control	300	2300	2000	18	30	+12	13	0	0	0.01	
1A	16	Control	300	2300	2000	24	31	+7	12	4	0	0.05	
1A	15	Control	300	2300	2000	27	36	+9	13	2	0	0.01	
1B	14	Control	300	2300	2000	18	30	+12	11	2	1	0.01	
1B	14	Control	300	2300	2000	22	29	+7	10	3	1	0.05	
2B	13	Control	300	2300	2000	16	30	+14	12	1	0	0.01	
2B	11	Control	300	2300	2000	20	29	+9	10	1	0	0.01	
4	15	Control	300	2300	2000	24	32	+8	10	3	2	0.05	
4	15	Control	300	2300	2000	22	29	+7	9	5	1	—	
5	12	Control	300	2300	2000	15	25	+10	9	3	0	—(0.05)	
5	12	Control	300	2300	2000	10	21	+11	11	1	0	0.01	
6	12	Control	300	2300	2000	12	23	+11	9	3	0	—(0.05)	
6	11	Control	300	2300	2000	12	21	+9	9	2	0	0.05	

[a] This experiment was significant if the experiments with very high preplayback rates and two unusual drops to zero are deleted. All significance levels in parentheses are of this nature (experiments 8, 9, and two controls), and if no significance level is stated it was not significant.

[b] All the multiple-frequency calls were produced so that the second and succeeding frequencies were 6 db down from the one preceding it unless otherwise stated. However, due to generation of harmonics when the fundamental was 90 Hz, the relationships were changed somewhat. See *Results* for information.

[c] Essentially instantaneous rise and decay of 10 μsec producing audible clicks. All other artificially produced calls have a rise and decay time of about 20 msec without obvious clicks.

[d] Played out louder than previous experiment (about 11–13 db above the other playbacks).

[e] Frequencies put together so that there was an unusual phase relationship compared to boatwhistle call.

[f] Normal growl of fish (rapid series of grunts) which we organized into the standard stimulatory organization (duration 300 msec, period 2300 msec).

[g] The leading edge of a real boatwhistle recorded on tape was cut off to make the short call.

[h] A continuous tone with a 12 db higher intermittent signal placed on the continuous tone.

[i] In this case (experiment 43) every other signal was down 6 db, then in the following two experiments the lower amplitude was used at two rates. In experiment 46 the signals were arranged so that the second was down 5 db and the third down 5 db from the second, so that three signals regularly alternated in amplitude; in the following two experiments the lower amplitude was played at stimulatory rates.

[j] The controls are tape-recorded copies of fish-produced boatwhistles played at the standardized stimulatory rate. The first was randomly ordered for each replicate set of a series, while the second was played at the end of each set.

so that he could not increase his rate. The analyses considered this problem where necessary by deleting high-rate fish.

Sound levels were such that the playback sound peaked the VU meter at −10 to −7, and the fish's call peaked it at 0 to +2 except in a few cases where the sound was made louder and peaked the meter at −1 to +1. The usual level represented the amplitude of a sound obtainable from a fish about 8–10 ft away.

Two sounds had essentially instantaneous rise times of 10 μsec, producing on and off clicks, whereas most other sounds were essentially clickless with rise and decay times of 25 msec.

Replicates, intended to be 16 for each type of playback, varied from 11 to 16 due to various failures of the system. One-tailed sign tests were used to determine whether playbacks were stimulatory.

In a different set of experiments, toadfish were centrally released in a square fish pen to see if they responded to boatwhistle playbacks from speakers located in the corners of the pen. These tests were carried out during June of 1961, 1962, 1963, and 1964. The pen was built beside the dock at Solomons Island in 2–4 ft of water. Figure 1 illustrates the experimental setup. Pairs of spawning fish served as the subjects. They were removed from

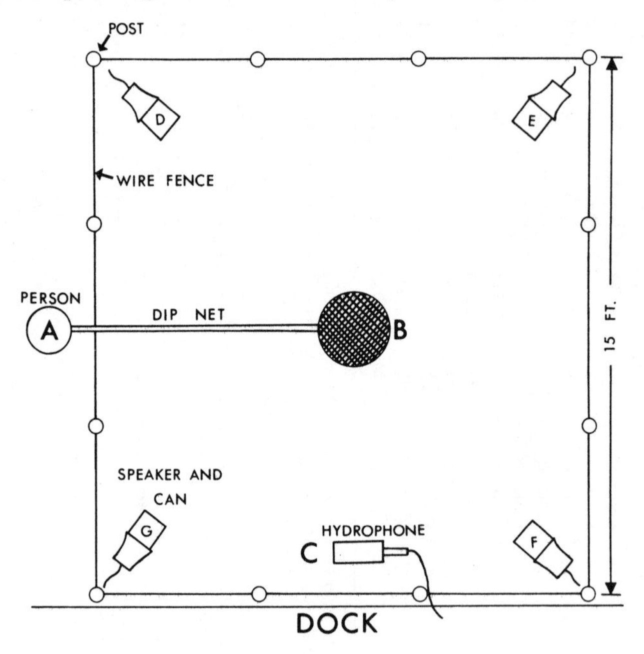

Fig. 1. Diagrammatic representation of experiments where male and female toadfish were introduced into the center of the pens to see if they reacted to either of two cans attached to speakers emanating sound. Two other cans acted as controls.

their nests and carried in a bucket to a platform near the pen. Each member of the pair was individually placed in a dip net, carried to the side of the pen, and lowered in the center of the pen where it was gently released. The fish's movements were then traced for up to 5 min. At each corner of the pen (Fig. 1) there was an open gallon can attached to a University MM21 speaker. Wires from all speakers led to the dock. From two of the four speakers, prerecorded natural boatwhistle calls were played out of a Message Repeater at a rate of 18 calls/min. All possible position combinations of speaker pairs were used at least ten times during the time of day selected for experiments. The amplitude represented a fish 8–10 ft away. Females were considered ripe if they contained more than ten ripe eggs when dissected after a test.

III. RESULTS

A. Physical Parameters for Facilitation

The 54 experimentally modified playbacks and 14 control playbacks indicated which parameters of the boatwhistle call elicited vocal facilitation (increased rates of calling) from the male (Table I). A standard set of conditions which caused increased calling was used for comparison: a sound duration of 300 msec and a period of about 2300 msec (about 22 calls/min). The control playbacks, consisting of fish-produced boatwhistles, caused an increase in boatwhistle calls (experiments, end of Table I). It did not matter whether the control test was randomly interspersed among the tests of a series or at the end of the tests (each playback of a series) played to one fish. The latter demonstrated that the fish continued to respond after as many as 11 consecutive experiments.

In experiments 1–25, the frequency (spectral content) of the sound was manipulated. The duration and period were more-or-less constant and of a stimulatory nature. Clearly, pure tones of 200, 300, and 400 Hz caused an increase in calling (see also experiments 43–54). Similar results were obtained with 180 Hz as the fundamental with various harmonics each 6 db down from the lower frequency (experiments 6–9). The purity of the spectral content is illustrated in Fig. 2. Experiments 10 and 11, with fast rise times and an audible click, did not alter the results. Higher frequencies of 600 and 800 Hz did not stimulate the fish. With only two exceptions, described below, sounds that contained frequencies of 90–100 Hz, close to low-frequency cutoff of the speakers, were not stimulatory (experiments 12–25). Experiment 12, using 50 Hz, actually was 100 Hz, the second harmonic of the sound generated by the speaker. It also produced weak harmonics when 90 and 100 Hz were utilized (experiments 12–15, Fig. 2). When 90 Hz was combined

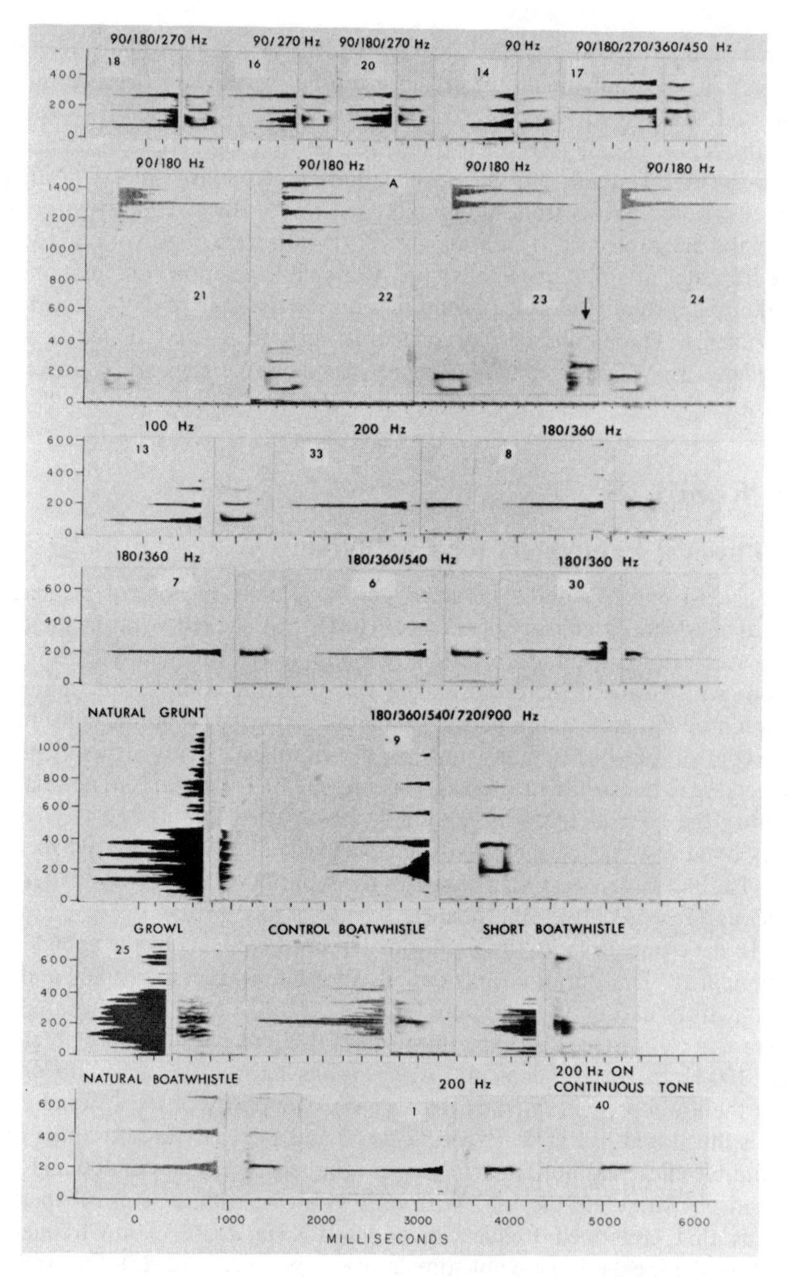

Fig. 2. Sonagraphic illustrations of the sounds played back to toadfish in the variously numbered experiments. In lines 1 and 3–7 a section is reproduced to the left of each signal (there is no section for the continuous tone, experiment 40). In line 2 the sections are above, and the frequency scale does not apply.

with other frequencies 6 db down, the sound generated underwater had about as much energy at 180 Hz as at 90 Hz due to the speaker generating 180 Hz. This was added to what was already in the original signal (experiments 17–20). No stimulation resulted. The original artificial signal had 180 Hz down 6 db from 90 Hz in experiments 21, 23, and 24, and 180 Hz and 90 Hz of equal amplitudes in experiment 22. However, the relationships were changed in the underwater signal due to the generation of harmonics in the speaker by the 90 Hz signal. In experiment 21, with a 90/180 Hz signal, 180 Hz was slightly greater in amplitude than 90 Hz but there was a strong 120 Hz component of unknown origin. This was probably the second harmonic of a 60 Hz hum. The 60 Hz can be heard by the fish (unpublished experiments). Thus low frequencies contained as much (or more) energy as 180 Hz, and stimulation did not take place. In experiments 22 and 23 the tones were purer, so that 180 Hz was several decibels greater than 90 Hz and stimulation then occurred. In experiment 24, using the signal of experiment 23, the sounds were played out much louder, imitating the amplitude of a fish about 1 ft from the experimental animal. The tones were less pure (Fig. 2) and did not cause an increase in calling. In experiments 18 and 19 tones were placed out of phase but also contained low-frequency energy. Thus there was no stimulation. A natural fish growl (experiment 25) was also not stimulatory.

When the sound duration was decreased to 75 msec and the standard period maintained, no increase in calling rate occurred (experiments 26–28). In fact, in 32 experiments call rate decreased, while in 13 it increased and in two it remained the same. A natural boatwhistle, shortened to 75 msec, did not induce increased calling. In most other cases where the duration was short (150 and 110 msec), but the period halved (thus rate doubled), there was no stimulation (experiments 29–32 and 34). Experiment 33 was supposed to have had a short duration but did not, and stimulation resulted. However, experiments 35 and 52 employed a pure 200-Hz tone of 75 and 150 msec duration, respectively, and a short period. In both cases calling increased. This did not happen when mixed tones and a 400-Hz tone were used (experiments 29–32 and 34). The results also suggest that a pure tone of 200 Hz is more stimulatory than mixed tones, particularly when 180 Hz is used as the fundamental.

Continuous tones markedly reduced calling or almost completely suppressed it (experiments 36–39). If a stimulatory signal 12 db higher was combined with a continuous tone, it also caused reduced calling (experiments 40–42).

A series of experiments (43–48) was performed where two and three stimulatory signals, differing in amplitude by 6 and 5 db, respectively, were utilized at rates of about 12 and 24/min. Then the lower-amplitude signals of the above calls were played out at two rates to make sure the toadfish heard the less intense signals and to show that the slower one was not stimulatory. All of them increased calling by the toadfish. The call rate of any one of the sounds in the three-amplitude experiment did not stimulate (8 calls/min).

A series of experiments (49–54) was performed with a constant period, an increased duration, and a decreased interval. Then the interval was kept constant, with the duration and period increasing. Stimulatory periods were used in all cases. Experiment 52 had a duration of 150 msec instead of the 300 msec that it was supposed to have, but experiment 33 nearly duplicated the necessary conditions. All caused an increase in calling except the one where the interval (200 msec) was too short to allow many calls. The details of the relation of boatwhistles to the interval are presented by Fish (this volume).

B. Pen Tests

The results of pen experiments with free-swimming males and females presented with playbacks of boatwhistle sounds are given in Table II. No females and only one male responded to a speaker in the limited number of tests made between 0900 and 1730 hr. The fact that a few females responded after 1700 hr caused me to restrict the majority of tests to 1700–2015 hr. Chi-square tests demonstrated that the responses of ripe females to speakers emitting sound were significantly different from their reactions to control (no sound) speakers (0.01 level). Also, ripe females responded to the sound can significantly more than males (0.01 level) or spent females. A response was scored when the fish entered the can or leaned against it; most entered the can (Table II). Males and spent females did not respond to sound cans, although the small number of residual male responses (four) may have some significance. Cans without sound were ignored in all but one case. Generally, the fish which responded went in a reasonably straight line to a can and sometimes had to turn in the center of the pen to go toward a sound source. However, other fish frequently went to the edge of the pen or toward a control can in essentially a straight line. The data were analyzed in two other ways to look for directional tendencies. How many ripe females and males went directly toward a sound can? Fifteen males did and 66 did not go in a straight line to a sound source, while 21 females did and 51 did not. These data include some tests not found in Table II. Eliminating the case where the sound cans were diagonal to each other, the pen was divided in half to see if there was a tendency to go in the general direction of the sound cans. Thus 38 males went toward the sound cans and 41 went the other way, whereas 37 females went toward the sound cans and 34 went away.

Table II. Results of Pen Tests with Male and Female Toadfish[a]

Time	Ripe female			Spent female			Ripe male		
	Sound can	Control can	No can	Sound can	Control can	No can	Sound can	Control can	No can
0900–1730	0	0	8	0	0	1	1	0	17
1730–2015	15[b]	1	28	0	0	18	4[c]	0	62

[a] Boatwhistle calls emanated from sound cans, while cans without sound acted as controls.
[b] Ten entered can.
[c] Two entered can.

IV. DISCUSSION

A. Signal Parameters

In any discussion of a communication system, a knowledge of the sensory capacity of the animal involved is vital. In recent years such knowledge has been obtained for some fishes, but due to the adaptive variation of the acoustico-lateralis system, each type of variant needs to be studied. Some of the main acoustic capacities of fishes will be mentioned here without any attempt to be comprehensive or historically complete.

The physical features of a boatwhistle that result in vocal facilitation are in general quite clear. First, a sound that has more energy at or near 100 Hz than it does at higher frequencies does not stimulate. Our experiments to date did not include pure low-frequency calls. In the cases where we precisely controlled intensity, there was an amplitude relation between low and high frequencies. Loud calls, equivalent to the call of a fish less than 1 ft away, caused a reduction in calling, whereas a call lowered in amplitude, simulating a fish 10 ft away, caused an increase in boatwhistling. Higher frequencies (180–400 Hz) were a positive stimulus. Although 600 and 800 Hz tones did not stimulate, it may be that the sounds were not heard. These frequencies are on the steeply rising side of the threshold curve, and high-intensity stimulation is thus called for (Fish and Offutt, in preparation). The purity of the sound (single frequency) used for stimulation did not seem to matter as long as it met two conditions: (1) there was no appreciable energy at low frequencies and (2) there was enough energy in the middle frequencies so the fish could hear the sound.

Duration, interval, and period, or rate, are interlocked aspects of the sound. For example, if duration is changed and not the period, then the interval is modified. Within certain limits, the duration is a factor in stimulating toadfish. Compared to a standard stimulatory duration of 300 msec, which is the approximate average duration of natural boatwhistle calls at Solomons Island, Md., shortened durations of 75 and 150 msec resulted in a loss of stimulatory value. In fact, a reduction in calling was obtained. Seventy-five milliseconds is roughly comparable to the duration of the natural grunt sound. In most of the above experiments the period was maintained at the standard, but in some cases the period was halved (or rate increased) to a level well above the natural highest rates of calling. Whereas most did not stimulate, two experiments at 200 Hz, (75 and 150 msec) resulted in a maximum stimulation.

Thus several problems need to be resolved with further playbacks. One is presented in the above experiments where high-rate, shortened calls of 200 Hz caused an increase in calling, but those of complex frequencies did not. Based on all of our other experiments, this should not have been the case, unless 200 Hz is more stimulatory than other frequencies and there is some aspect of energy input in the normal calls that is not clear in our experiments. This finding may be related to on–off auditory neuron stimulation at restricted frequencies. The amplitude relationships of low- and high-frequency signals need further study with a speaker that does not generate harmonics, and pure low-frequency playbacks are needed. Absolute sound levels must also be obtained.

We are now in a position to state clearly some of the properties of the boatwhistle call, a signal of the harmonic-frequency type, that evoke a response in the toadfish. There are several important properties which were not expected when a theoretical classification of fish sounds was proposed (Winn, 1964) and the important properties for information transfer were hypothesized.

Duration is important over a broad range; if too short or too long, no response occurs. An intermittent signal is required, loud enough for the fish to hear, but not too loud. The rate of calling must be above a threshold rate, with an interval of sufficient duration to allow for an increased response. The fundamental must be around 200–400 Hz, with no energy exceeding these frequencies in the 100 Hz range. Thus, spectral amplitude relations are also important.

In other studies, several investigators (for example, see Myrberg *et al.*, 1969; Myrberg, this volume) have shown that sharks were attracted to irregularly handpulsed harmonic sounds (overdriven) with fundamentals varying from 55 to 500 Hz. Pulse intervals varied from 0.05 to 1 sec. Various broad-band filtered sounds (25–50, 50–100, 150–300, 400–800, and 500–1000 Hz), irregularly pulsed, also attracted sharks but perhaps to a lesser extent. Pure sine waves, also irregularly pulsed, did not attract them. Although the pulsing was important, no controlled tests on different pulse rates were made. However, the results firmly established that multiple-frequency sounds were necessary for attraction and localization. Moulton (1956), after playing back sounds to sea robins, found that production of the staccato call could be initiated by imitations of the call and by the call itself, whereas suppression resulted with playbacks of 200–600 Hz sine waves.

Also, other aspects of pulse-rate parameters may be important for discrimination. Myrberg (personal communication, 1969) suggests that interspecific discrimination by certain pomacentrids may be based on temporal patterning of pulsed sounds similar in amplitude and frequency. Jasper and Littlejohn (1971) demonstrated in controlled experiments that two species of frogs differentiate each other's calls on the basis of pulse repetition rate or a related parameter.

It appears that a sound source must contain multiple frequencies and be of an intermittent nature to be adequate for localization. The natural boatwhistle call of the toadfish is of this type, as would be the thrashing of struggling fish that presumably attracts sharks. This evidence suggests that multiple neural input is necessary, either through differentially responding auditory neurons or through the lateral line neurons or perhaps both.

Single auditory neurons of *Cottus scorpius* were classified into four units (Enger, 1963) with regard to their spontaneous activity: irregular, burst, regular, and not spontaneously active. Thus, within certain limits, frequency could be detected by the volley principle. The spectral sensitivity and threshold for the various types also varied. Some units responded up to 200 Hz and some to 300–500 Hz. Thus some peripheral analyzer is involved. In more highly evolved teleosts an even more sophisticated system might be present. A simple system like the one above, with a change in neuronal sensitivity in the two frequency bands, could account for the toadfish response.

Several systems can be hypothesized to explain the fact that the male toadfish responds to a signal of 200–400 Hz, but does not increase his calling rate when sufficient amplitudes of the lower frequencies are added. The low frequency could enter the lateral line and inhibit the response, or it could activate certain low-frequency responding neurons in the ear that inhibit a response. Perhaps the reverse is true, and the high frequency rather than stimulating merely releases an inhibitory system, and the low frequency prevents this release. Regardless of the system employed, it now will be important to make a neurophysiological study of these mechanisms. The toadfish system is not too different from that of the bullfrog (Capranica, 1965, 1966) but appears slightly simpler. A bullfrog will respond with a mating call when in a playback call there is reasonably high energy in a low- and high-frequency region but much less energy in a midfrequency region. An optimum repetition rate exists. More graded responses occur than seem to exist in the toadfish, although optimum responses at 200 Hz suggest the possibility of some grading.

Ambient underwater sound contains many changing and some constant components. Fish are highly tuned to this background noise, and they can detect certain types of unfamiliar changes in it. A few irregularly produced sounds, unless very loud, do not seem to have any effect. Background noise playbacks reduced toadfish calling to some extent (Winn, 1967). In an unpublished experiment where toadfish were transferred to a 10,000 gal tank in Narragansett, R.I., they called for over 6 weeks. Although walking on the cement near the tank had no effect, we found that merely leaning against it stopped their calling. The experiments described earlier demonstrated that continuous tones markedly reduced calling. Even when we placed higher-amplitude intermittent signals of the same frequency (that alone are stimulatory) on a continuous tone, suppression was obtained. Therefore, 60 cycle sound (suppressed calling in unpublished experiments) and other background noise must be carefully controlled in playback experiments. Even a stimulatory signal in the presence of high background noise might produce an opposite result. This fine tuning to the background may have significance other than that of allowing the toadfish to select signals out of it.

B. Signal Systems

The sounds produced by a wide variety of fishes are organized in a simple fashion, but only in a few cases do we have any idea about their message content. There seems to be a lack of a finely graded series of signals, as seen in birds, or a subdivision into many kinds and degrees of responses. However, our knowledge is in a primitive state, and future research may indicate otherwise. Due to the rarity of more than transitory pairing in fishes, we would never expect the development of many kinds of acoustic signals associated with long pairing and nesting bonds found in, for instance, birds.

The properties of acoustic signals in some fishes may be categorized in five basic ways (Fig. 3). The first and second are variable-time-interval and fixed-time-interval signals. The time between any unit which is heard and viewed as one sound on a spectrogram (grunt, knock, etc.) may be variable or fixed. This does not include pulses which are smaller parts of a sound unit produced by a muscle–swim bladder mechanism. These pulses can be resolved by slower playback and are the result of single muscle contractions (e.g., Winn and Marshall, 1963). If the sound is produced by an air bladder mecha-

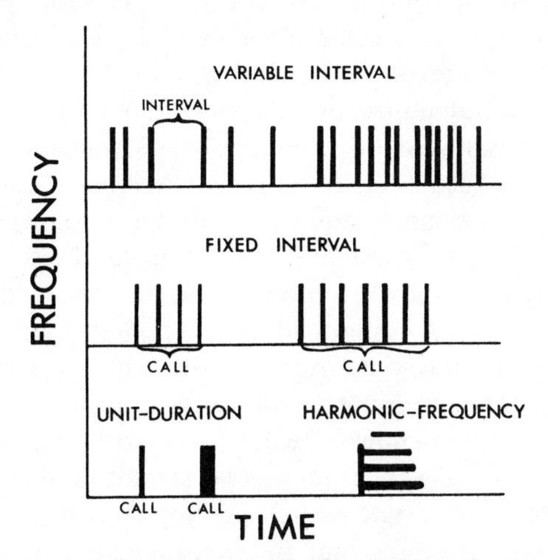

Fig. 3. Diagrammatic representation of the organization of fish sonic signals. [Modified and reprinted with permission from H. E. Winn, The biological significance of fish sounds, *in* "Marine Bio-Acoustics" (W. N. Tavolga, ed.) Pergamon Press.]

nism, the variable- and fixed-interval calls are usually nonharmonic sounds containing frequencies from 2000–3000 Hz down to frequencies below 100 Hz, with the greatest amplitude in the lower frequencies. Sounds produced by pharyngeal teeth and other stridulatory sounds may have the greatest amplitude distributed more uniformly throughout the various frequencies, and can range to well over 6000 Hz. In the third and fourth types of signals, the duration of any unit of sound is lengthened (unit-duration signal) or the amount of time during which units are produced is varied (time-length signal). The first is uncommon as a single feature, and the second is apparently frequently done with fixed-interval sounds. Harmonic-frequency signals of longer duration comprise the fifth category. In some fishes at least the variable-time-interval type can be graded by varying the interval, while the fixed-interval and harmonic-frequency types can be graded by varying the length of time they are emitted. It now seems likely that fish can make use of variations in amplitude to communicate information. We can expect intermediates among the above basic types. The various kinds of signals are schematically shown in Fig. 3.

Another type of classification is based on the proximate cause for production of a sound. Most calls seem to require general or specific changes in environmental stimuli such as the grunt of the toadfish. These would be externally driven sounds even though the fish must be in a proper physiological state. The other type occurs when the animal is in the required environmental and physiological state and "spontaneously" produces sounds, i.e., internally driven sounds. The only well-known call of this type is the "boatwhistle" call of *Opsanus tau* and other toadfishes. Serranids, bagrids, sciaenids, prionotids, and some other fishes have endogenously driven signals, but none have been studied in any detail.

Different species utilize the above types to variable degrees. A large number of species known to produce sounds will not be considered here, but rather I shall concentrate on those that are known in more detail.

One of the better understood is the oyster toadfish, *Opsanus tau*, which emits grunt sounds as variable-interval signals. These may grade into a fixed-interval-type signal when produced at the maximum rate (growls, Fig. 2). The other call of the oyster toadfish is "spontaneous" (i.e., given by males in the nest), and is a harmonic-frequency signal of long but relatively constant duration. The other species of toadfishes seem to have a similar repertoire.

From what little is known, the northern midshipman (*Porichthys notatus*) emits variable-interval and fixed-interval sounds, and possibly a harmonic-frequency sound that is not as well developed as in the toadfishes (Cohen and Winn, 1967). It seems that the midshipman does not have as highly developed a system as *Opsanus*. There is doubt whether some sounds of catfishes (Tavolga, 1960), *Myrpristis jacobus*, piranhas (Winn, unpublished observations), and other fishes (Fish, 1954), which have weak harmonics close together, should be classified as biologically significant harmonic sounds. The sounds are intermediate in development, and as a first thought it appears that the harmonic content of the sound is unimportant in the communication of information. Future research may well show otherwise. *Bagre marinus* has two variable-interval sounds, a low-pitched grunt and a high-pitched yelp, plus a harmonic-frequency sound, the "sob" (Tavolga, 1960), but the harmonics do not originate in the air bladder (Tavolga, 1962). *Galeichthys felis* produces long (100 msec or over) grunts under duress and short grunts when in the company of other individuals of the same species. Here is the first example of two sounds differing primarily in duration of the unit (a grunt). The satinfin shiner has variable-interval and fixed-interval types of sound produced in different contexts (Stout, 1963*a,b*). *Holocentrus rufus* has a variable-interval gruntlike signal and an entirely distinct fixed-interval staccato call. Here, there is no gradation of one into the other. The length of staccato call depends on the strength of the stimulus. The captive black grouper, *Mycteroperca bonaci*, emits fixed-interval sounds with about five beats in each group (Tavolga, 1960). The reef squirrelfish, *Holocentrus coruscus*, has only a variable-interval signal which is graded according to the temporal and strength aspects of the stimulus (Winn, unpublished). These sounds were recorded under several conditions, particularly during territorial behavior, but never during reproductive behavior, when it is possible that other types of sounds are emitted. The croaking gourami, *Trichopsis vittatus*, seems to produce only variable-interval sounds (Marshall, work in progress). Tavolga (1956) stated that *Bathygobius soporator* gives variable-interval signals and that the interval decreased as a courting male and female increased the vigor and amplitude of courtship. This may grade into an occasional fixed-interval-type signal, as Tavolga noted that under maximum stimulation, sounds are sometimes given in quick succession of fours or fives.

Further characterization of fish signals was made by Fish and Mowbray (1970), who represented most of the known sounds in the fishes of the Western North Atlantic. They characterized fish pulse-type sounds as represented in Table III. One can see that fishes produce short and medium length single multifrequency sounds. These can be repeated over longer periods both with fixed and variable intervals. Multipulse gruntlike signals are also produced at slow rates. Once the pulsing is repeated very rapidly at a constant rate of usually longer duration, one obtains harmonic sounds with higher fundamentals. These are all muscle–swim bladder mechanisms, whereas high-frequency stridulatory sounds are produced by rubbing teeth or bones together.

When sounds are categorized into variable-interval, fixed-interval,

Table III. Typical Characteristics of the Underwater Sounds of Fishes (from Fish and Mowbray, 1970)

Sound type	Basic sound	Duration[a]	Repetitive sound form	Repetitive duration[a]	Mechanism	Fundamental frequency[a,b]
Pulse	Thump, boom	20–90 msec (avg. 40)	Rumble, growl, drum	50 msec to 5 sec	Swim bladder	40–200 Hz
Pulse	Knock, escape	2–20 msec (avg. 10)	Croak, honk	30 msec to 4 sec	Swim bladder	150–750 Hz
Multiple pulse	Grunt, cluck, bark	20–120 msec (avg. 60)	Burst of basic sounds	50 msec to 5 sec	Swim bladder	50–600 Hz
Pulse	Click, snap	1 msec	Scrape, rasp	2–150 msec	Stridulation	1–5 kHz
Multiple pulse	Grate, scratchy grunt	15–200 msec	Burst of basic sounds	50 msec to 5 sec	Swim bladder, stridulation	150–750 Hz
Sustained	Groan, growl, honk	100 msec to 2 sec	—	—	Swim bladder	50–200 Hz
Sustained	Groan, growl	100 msec to 3 sec	—	—	Skeletal muscles	40–80 Hz

[a] Values indicated are representative and not necessarily absolute limits. Averages are for the typical sounds illustrated by oscillograms.
[b] Fundamental frequencies of pulses are those measured from oscillograms and are not repetition rates.

unit-duration, time-length, and harmonic-frequency signals, several important principles become apparent. First, most if not all sounds are obviously different in their temporal patterning. This may be the feature that communicates intraspecific information during courtship, territorial defense, and reactions to predators. Information can be graded by varying the interval and length of time that units are emitted. Amplitude and frequency characteristics of most acoustic signals may also be important. This is seemingly true of *Bathygobius soporator*, *Holocentrus coruscus*, grunting *Opsanus tau*, and *Trichopsis vittatus*. It must be cautioned that considerable experimental evidence is necessary. Tavolga (1958) demonstrated that positive approach by males of *B. soporator* could be induced by artificial sounds such as pure sine waves and other sounds that differed from the natural sounds (positive responses to increased sound levels of a hundredfold, frequencies from 100 to 300 Hz, and pulse durations from 75 to 150 msec). Moulton (1956) obtained sound mimetic behavior with sounds of 200–600 Hz. Stout (1963*a*,*b*) obtained differential responses of males and females of *Notropis analostanus*. Myrberg (this volume) cites other instances of the effect of playbacks on behavior, especially with the bicolor damselfish.

It has been shown that *Phoximus laevis*, a cypriniform, can discriminate between narrow frequencies (*cf.* review by Dijkgraaf, 1960). It may tentatively be presumed that *N. analostanus* has this ability, but its acoustic signals are variable- and fixed-interval-type sounds, which would imply that only temporal patterning is used for discrimination. Frequency and amplitude characteristics might well be used to differentiate its own sounds from other sounds of the environment. Schneider and Hasler (1960) suggested that repetition rate might be the important factor for species recognition in the Sciaenidae.

The harmonic type of call is uncommon. The best known one is the boatwhistle call of the male oyster toadfish. Here the primary features of the call are its frequency characteristics and its long but fixed duration, when compared to grunts. It has now been demonstrated that repetition rate, intervals, duration, frequency, and even amplitude relations are all important parameters resulting in vocal responses to playbacks. Thus considerable redundancy is built into the system.

Little is known about the patterning of the high-frequency stridulatory sounds. These tend to grade from variable-interval into fixed-interval types. Their general value may be in eliciting slight escape reactions by nearby fish when one is attacked by a predator or in the presence of another alarming stimulus. Specific variations in patterning may be less significant here because there may be no advantage for discrimination. In fact, it could be of survival value to lack discrimination.

It has been assumed by investigators studying fish sounds that hearing was the sense that extended a fish's world beyond its limited visual sense. This may be true when the visual range is limited to a few centimeters, but usually the water's clarity allows for adequate vision over at least several meters in the surface layers of the oceans. The advantages of sound are usually stated to be the high rate of conduction, roughly five times that in air, and the low rate of attenuation in the lower frequencies. However, it appears that these advantages present difficulties. Localization appears to be difficult, and hearing is limited to a narrow low-frequency range. Fish (1969) determined thresholds of response to continuous tones and normal boatwhistles in the toadfish. He established that a toadfish could respond to a boatwhistle call of another toadfish only if they were within 3–4 m of each other. At the moment, these thresholds are believed to be close to true physiological thresholds. The boatwhistle call is one of the loudest fish sounds, about 50 db, re 1 μbar. In fish with less intense calls, the range would be even more reduced unless hearing was improved, as in ostariophysids and holocentrids. However, in order to overcome the disadvantages of sound in water, a large animal is needed with acoustic isolation of each ear. The apparent great advantages of the acoustic transmission channel bring about serious disadvantages that severely limit its refined use in fishes. Based on the available evidence, the acoustic world of small fishes is extremely limited in range in comparison to that of air-dwelling vertebrates. However, sharks seem to be able to localize sound sources over 25 m away (Myrberg *et al.*, 1969).

We see in the toadfish the presence of two basic sounds, one type of which is produced by most vertebrates and is referred to as "song" in birds (Winn, 1964). Some of the characteristics of song (Thorpe, 1961) are as follows: it is related to reproduction and territorial behavior; it is long and made up of a particular pattern of notes; it must be a type to allow for localization; it is given for long periods at regular intervals; it is generally internally driven ("spontaneous"); and it is frequently produced by unmated males early in the reproductive season. Earlier work (Gray and Winn, 1961; Winn, 1964, 1967) has shown that the boatwhistle call of the toadfish fits all these characteristics at a primitive level, except that we know very little about the characteristics of a fish call that will permit localization. In fact, localization has been demonstrated in only a few cases. However, in these cases at least a multiple-frequency sound is suspected.

The ability of the ostariophysine goldfish to discriminate frequencies and intensity has been studied (Jacobs and Tavolga, 1967, 1968). Goldfish could discriminate an intensity difference of 3–6 db in their normal range of hearing, depending upon frequency, and could discriminate between 9.4 Hz (4.7%) and 17.4 Hz (3.5%) at 200 and 500 Hz, respectively. Dijkgraaf (1952) showed that three nonostariophysine fishes were capable of discriminating only a 9–10% difference. It is clear that fish can discriminate these various differences, but it has not been shown how this ability is utilized in their natural behavior. As Jacobs and Tavolga (1967) postulated, high difference limens would make it difficult to use the information for discrimination, but it would be useful for signal detection. The prime problem is that natural behavioral discrimination studies have not been made. In this chapter it has been demonstrated that the nonostariophysine toadfish can use frequency and amplitude for discrimination. There is also a different response shown to sounds with high- and low-frequency components, depending on the relative amplitude of the two components. In some cases the amplitude of the total signal has different effects—stimulation or suppression.

Tavolga (1968) looked at the relative development of levels of interaction in animal communication. He classified them into vegetative, tonic, phasic, signal, symbolic, and language levels. Our only concern here is the phasic level, which involves discontinuous, more-or-less regular changes or events in the development of the organization of animals. These usually involve broad and multichannel stimuli with some specialization on the part of the emitter and receiver. The next level, signal, involves emitters with special structures producing stimuli along a single channel, usually in a narrow band. Tavolga (1968) assumes that the majority of sounds produced by swim bladder mechanisms in fishes range from 50 to 100 Hz, but there are a significant number of exceptions to this (see Fish and Mowbray, 1970). Finally, Tavolga concludes that, based on the present state of knowledge, "until such experimental knowledge is available, we must assume that with the possible exception of the toadfish, these acoustic interactions among fishes are primarily on the phasic level." The experiments described here and those of Fish (this volume) make it unquestionable that the toadfish system has developed to the signal level, meeting all requirements of the definition. Furthermore, I expect that many other fishes will meet these requirements once they are studied adequately.

It has also been tacitly assumed that most fishes have only one or two discrete signals. The toadfish has two, one of which can be continuously graded according to the intensity of an agonistic encounter. However, there is a strong suggestion that some squirrelfishes have at least five signal types (Salmon, 1967; Winn, unpublished). This is without knowing anything about sounds related to reproduction. Myrberg (this volume) reports on what appear to be five discrete signals produced by the bicolor damselfish.

The sound system described here is highly adaptive in the toadfish's normal environment. Two sounds, the grunt and boatwhistle (Fig. 3), are produced, one in aggressive encounters particularly by males and the other "spontaneously" from the nest cavity only by males. The grunt is a continuous variable-interval-type signal which increases in rate as any encounter becomes more intense. It can increase in rate until the single grunts are produced almost continuously and the call is better characterized as a growl. The two signals are differentiated by their duration, frequency content, and rate. I have seldom heard other sounds similar to the boatwhistle around natural populations of toadfish. Thus there is no known environmental situation such as a closely related sound that requires further specialization of the boatwhistle call. There are occasionally situations where other fish which spawn in bays along the Atlantic Coast produce gruntlike sounds near toadfish, but in general they would have no effect on the boatwhistle call. The utilization of grunts is most frequent when two male toadfish are close and in visual contact so that grunts produced several or more feet away do not interfere with the signal–response system.

We are a long way from fully understanding the message content of the two basic sounds of the toadfish, the grunt and boatwhistle, other than by analogy to similar vertebrate acoustic systems. It is even possible that there are more than two discrete signals, although this seems unlikely.

There are many possibilities as to the function of the boatwhistle in toadfish. By analogy to other vertebrates where a male calls early in the season from a territory enclosing the nest site, the call could attract females, stimulate them physiologically, stimulate adjacent males, maintain territories, etc. Such a call has, as a minimum, the message of the presence of a male in reproductive condition. Whether it is read that way or not is another matter. Our studies (Fish, this volume; Winn, 1964, 1967) have adequately documented that one message is to "increase rate" of calling when a high-rate call is heard. One difficulty was that most males are not calling at the minimum stimulatory rate of 12 calls/min. Average calling in the controls of Table I of this chapter was about 7 calls/min. Thus most animals are usually calling below the stimulatory rate. However, several experiments were designed to test if two, three, or more animals calling would make the composite signal stimulatory. In general, there is no overlap of calls (Winn, 1967; Fish, this volume). The experiments with two and three signals of different amplitude did stimulate a male to call faster, so that there must be a group effect. An isolated animal would not receive this stimulation. Although we have not been able to test this experimentally to date, one hypothesis of the function of the song is that it aids in maintaining a physiological state of reproductive readiness in males. A small group of males would have a higher rate of reproductive success than a single one, and in fact a single male might be stimulated to attempt to seek the closeness of other males and thus encourage aggregation. The availability of nest sites would then control the realization of such an aggregation.

Evidence to date suggests that the boatwhistle call stimulates the female to move, perhaps toward the source of sound. The only time a male increases his rate of calling, except when other calls are heard above the minimum stimulatory rate, is when a ripe female comes within a few feet of the male (experimental results of Gray and Winn, 1961). Other disturbances either stop boatwhistling or have no effect on it. Fish (this volume) observed a male to increase his call when a female came near the nest. She then entered the tile. The results in this chapter demonstrate that calls played out from nest cans caused a significant number of females to enter the can, whereas spent females and males did not. Although the percentage reacting was not extremely high, the results seem reasonable. In fact, it is a wonder that any reacted at all under the experimental conditions. A pair interrupted in the middle of spawning was poured into a bucket from a tile, carried over to the test pen, placed one after the other into a net, and released in the center of the pen. There must have been a strong drive to react normally under these conditions. Thus to the female the message reads, "come to the signal and enter the nest cavity," and it need not be reinforced by the presence of a male in the nest. However, such reinforcement conceivably could have resulted in a higher response level. When the pen was divided into sections, there did not seem to be a tendency for either sex to go toward the cans in general. Thus we cannot say at this time whether the reaction of the females to the sound started at the released point or whether they swam toward a can by chance and the specific reaction appeared somewhere along the path to the can emanating sound. The straight lines of reacting females may argue for the former, but then many paths were more-or-less straight even if toward an empty side of the pen. As Fish (this volume) has pointed out, one male, when stimulated to call faster by a female, stimulates other nearby males to also call faster, thus increasing the chances of a female locating a male. This is also what happens in other animals with similar call systems such as insects, fiddler crabs, frogs, and birds.

It is known that males first fill up many tiles when set in a tight row but that after a week or so they become spaced several feet apart (Winn, unpublished observations). It is possible that boatwhistling aids in maintaining this territorial distance, because in our experiments where very loud calls were played back they tended to inhibit the male from calling for at least a short period. Further experiments are necessary to test this hypothesis.

ACKNOWLEDGMENTS

Of the many persons who contributed to this research I would like particularly to express my appreciation to Pierce Fenhagen and Raymond Kenney, who helped carry out the playback experiments and performed data reduction. Joseph Marshall helped with many of the pen test experiments. Robert Haas helped make some of the playback tapes. My thanks go to various personnel at the Natural Resources Institute, Chesapeake Biological Laboratory, Solomons, Md., for their help and for allowing me to use their facilities over the past 8 years.

Figures 1 and 3 were drawn by Lois Winn. Michael Salmon, Michael Fine, Lois Winn, and James Fish kindly reviewed the manuscript.

REFERENCES

Capranica, R. R., 1965, "The Evoked Vocal Response of the Bullfrog: A Study of Communication by Sound," The M. I. T. Press, Cambridge, Mass.

Capranica, R. R., 1966, Vocal response of the bullfrog to natural and synthetic mating calls, *J. Acoust. Soc. Am.* **40:** 1131–1139.

Cohen, M. J., and Winn, H. E., 1967, Electrophysiological observations on hearing and sound production in the fish, *Porichthys notatus*, *J. Exptl. Zool.* **165:** 355–370.

Dijkgraaf, S., 1952, Bau und Funktionen der Seitenorgane und des Ohrlabyrinths bei Fischen, *Experientia* **8:** 205–217.

Dijkgraaf, S., 1960, Hearing in bony fishes, *Proc. Roy. Soc. London Ser.* B **152:** 51–64.

Enger, P. S., 1963, Single unit activity in the peripheral auditory system of a teleost fish, *Acta Physiol. Scand.* **59** (Suppl. 210): 1–48.

Fish, J. F., 1969, The effect of sound playback on the toadfish (*Opsanus tau*), Ph. D. thesis, University of Rhode Island.

Fish, M. P., 1954, The character and significance of sound production among fishes of the Western North Atlantic, *Bull. Bingham Oceanogr. Coll.* **14:** 1–109.

Fish, M. P., and Mowbray, W. H., 1970, "Sounds of Western North Atlantic Fishes," Johns Hopkins Press, Baltimore.

Gray, G.-A., and Winn, H. E., 1961, Reproductive ecology and sound production of the toadfish, *Opsanus tau*, *Ecology* **42:** 274–282.

Jacobs, D. W., and Tavolga, W. N., 1967, Acoustic intensity limens in the goldfish, *Anim. Behav.* **15:** 324–335.

Jacobs, D. W., and Tavolga, W. N., 1968, Acoustic frequency discrimination in the goldfish, *Anim. Behav.* **16:** 67–71.

Jasper, J. L., and Littlejohn, M. J., 1971, Pulse repetition rate as the basis for mating call discrimination by two sympatric species of *Hyla*, *Copiea* **1971:** 154–156.

Moulton, J. M., 1956, Influencing the calling of sea robins (*Prinotus* spp.) with sound, *Biol. Bull.* **111:** 393–398.

Myrberg, A. A., Jr., Banner, A., and Richard, J. D., 1969, Shark attraction using a video-acoustic system, *Mar. Biol.* **2:** 264–276.

Salmon, M., 1967, Acoustical behavior of the menpachi, *Myripristis berndti*, in Hawaii, *Pac. Sci.* **21:** 364–381.

Schneider, H., and Hasler, A. D., 1960, Laute und Lauterzeugung beim Süsserwassertrommler *Aplodinotus grunniens* Rafinesque (Sciaenidae, Pisces), *Z. vergl. Physiol.* **43:** 499–517.

Stout, J. F., 1963*a*, Sound communication during the reproductive behavior of *Notropis analostanus* (Pisces: Cyprinidae), Ph. D. thesis, University of Maryland.

Stout, J. F., 1963*b*, The significance of sound production during the reproductive behavior of *Notropis analostanus* (family Cyprinidae), *Anim. Behav.* **11:** 83–92.

Tavolga, W. N., 1956, Visual, chemical and sound stimuli as cues in the sex discriminatory behavior of the gobiid fish, *Bathygobius soporator*, *Zoologica* **41:** 49–64.

Tavolga, W. N., 1958, The significance of underwater sounds produced by males of the gobiid fish, *Bathygobius soporator*, *Physiol. Zool.* **31**: 259–271.

Tavolga, W. N., 1960, Sound production and underwater communication in fishes, *in* "Animal Sounds and Communication" (W. E. Lanyon and W. N. Tavolga, eds.) pp. 93–136, Publ. No. 7, Am. Inst. Biol. Sci., Washington, D. C.

Tavolga, W. N., 1962, Mechanisms of sound production in the ariid catfishes *Galeichthys* and *Bagre*, *Bull. Am. Mus. Nat. Hist.* **124**: 1–30.

Tavolga, W. N., 1967, Masking noise and auditory thresholds in fishes, *in* "Marine Bio-Acoustics" (W. N. Tavolga, ed.) Vol. 2, pp. 233–245, Pergamon Press, Oxford.

Tavolga, W. N., 1968, Fishes, *in* "Animal Communication" (T. A. Sebeok, ed.) pp. 271–288, Indiana University Press, Bloomington.

Thorpe, W. H., 1961, "Birdsong. The Biology of Vocal Communication and Expression in Birds," Cambridge University Press, Cambridge.

Winn, H. E., 1964, The biological significance of fish sounds, *in* "Marine Bio-Acoustics" (W. N. Tavolga, ed.) pp. 213–231, Pergamon Press, Oxford.

Winn, H. E., 1967, Vocal facilitation and the biological significance of toadfish sounds, *in* "Marine Bio-Acoustics" (W. N. Tavolga, ed.) Vol. 2, pp. 283–304, Pergamon Press, Oxford.

Winn, H. E., and Marshall, J. A., 1963, Sound-producing organ of the squirrelfish, *Holocentrus rufus*, *Physiol. Zool.* **36**: 34–44.

Chapter 11

THE EFFECT OF SOUND PLAYBACK ON THE TOADFISH*

James F. Fish

Naval Undersea Research and Development Center
San Diego, California

I. INTRODUCTION

Before we can fully understand sound communication in fishes, it is necessary to learn more about the communication of specific forms in their natural environments. This investigation was conducted to clarify some aspects of the breeding season sounds of the toadfish, *Opsanus tau* (Linnaeus).

Prerecorded natural toadfish sounds (boatwhistles) and various puretone frequencies were played back underwater to individual calling male fish on their breeding nests. Six different experiments were performed to try to establish what parameters of the fish sounds were most important for transferring information. The specific objectives of the experiments were (a) in the calling-rate experiment to determine how closely calling rate could be controlled by playback rate, (b) in the boatwhistle-pattern experiment to determine if calling pattern could be controlled and predicted by playback of certain boatwhistle patterns, (c) in the cycle-time experiment to determine if there was an optimum evoked-response calling rate, (d) in the antiphony experiment to determine if rhythmic alternation could be established between

* This research was supported partially by an NDEA Title IV fellowship to the author, and also by NONR contract 396(08), and USPHS research grant 5-R01-NB06397 to H. E. Winn. The report was submitted in partial fulfillment of the requirements for the author's Ph. D. at the University of Rhode Island.

a calling fish and a boatwhistle playback with the playback rate controlled by the response time of the fish, (e) in the tone-pattern experiment to determine if calling pattern could be controlled and predicted by playback of certain pure-tone patterns, and (f) in the response-time experiment to estimate the vocal response time to certain sounds.

Answers to these questions would help to evaluate the complexity of the toadfish vocal communication system. A well-developed system would be a real advantage to many marine animals and particularly to toadfish, which often breed in water where the visibility is only 1–2 m. An understanding of sound communication in fishes may enable us to control their movements and/or vocal behavior with sound. It is desirable to be able to attract commercially important species and guide them from dangerous areas, to repel undesirable forms, and to silence highly soniferous species which interfere with passive sonar systems.

This chapter examines the vocal response to sound playback rather than movement. Myrberg (this volume) discusses the effect of sound on movements of fish.

The toadfish was chosen as the experimental animal for several reasons. It is a slothful nesting species; individuals remain in the same location for several weeks. Consequently, single animals could be presented many different playbacks. Its life history has been well described by Gudger (1910) and others, and its reproductive ecology and sound production have been described by Gray and Winn (1961). In Chesapeake Bay in late May or early June, males move into shelters such as tin cans and jars, or make nests by burrowing under rocks; females enter the nests, deposit their eggs, and swim away leaving the males to guard them. During the breeding season, two sounds are produced by the nesting males, one referred to as a boatwhistle, produced spontaneously, and the other a grunt. Apparently, females make only the grunt. Several investigators have described these sounds (Fish, 1954; Tavolga, 1958*a*, 1965).

Winn (1967) showed that nesting male toadfish would respond vocally, by increasing their calling rates, to playbacks of prerecorded natural boatwhistle sounds when the repetition rate of the playback was 18 sounds per minute or greater; a rate of 10 sounds per minute was not stimulatory. High-level continuous tones suppressed calling, and 200-Hz tone bursts, with durations similar to natural boatwhistles, increased it (Winn, this volume). These important discoveries stimulated numerous questions that could only be answered by further playback experiments.

No experiments have been performed in the past to determine if fish can discriminate different sound patterns or if their calling patterns can be controlled by sound playback. There is little information on how well fish can discriminate rate, except that toadfish can distinguish fast from slow calling rates, as evidenced by Winn's playback experiments. Antiphonal calling in fish has not been experimentally tested, and there are no estimates for the vocal response time to sound playback or other stimuli.

The toadfish, *Opsanus tau*, is a unique fish for these detailed sound playback field experiments. Consequently, it is essential to learn everything we can about its vocal behavior. Results of these experiments may be useful in designing other playback experiments with different species of fish.

II. METHODS

A. Experimental Design

Sound playback experiments were conducted at the Chesapeake Biological Laboratory, Solomons, Maryland, in June 1967 and 1968. A 150-m-long pier provided an excellent work platform over water 0.5–3 m deep. Before the start of breeding season, empty tin cans, jars, and other obvious shelters were removed from the experimental area. Over one hundred 30-cm-long sewer pipes, with one end cemented closed, were then positioned on the bottom beside the dock pilings (2.5 m apart) with the openings facing away from the pier.

Shelters were inspected for the presence of adult fish, eggs, and young either by diving or by retrieving them with special tongs. Nearly every shelter contained a male fish by June 1. A tagging experiment, performed the year before the start of the sound playback experiments, confirmed that the same fish remained in the same shelters for nearly 2 months. For details of this experiment, see Fish (1969).

Each experiment discussed in this chapter utilized a different series of playbacks. The number of times each playback was presented is referred to in the text as "replicates." Boatwhistles are frequently called BW and grunts G.

The recording system consisted of a Hydro Product's R-130 hydrophone, a power supply and amplifier, and a Uher 4000-L tape recorder (system frequency response 50–10,000 Hz $\pm$ 3 db). Sounds were played back using another Uher 4000-L, connected to a Nagra portable amplifier driving a University underwater loudspeaker (system frequency response 100–2000 Hz $\pm$ 3 db). An oscilloscope was connected across the input of the recording system tape recorder to monitor the waveform of the projected sounds.

To record the playback sounds and the sounds produced by a test fish, the hydrophone was gently lowered to the bottom near the opening of a shelter. The sound pressure level of the calls from the test fish was at least 20 db above the sounds of the nearest neighbor 2.5 m away. To play sounds back to a fish, the underwater speaker was placed 1.5 m in front of a shelter and the playback level adjusted until it was 8–10 db below the fish's calls. This equipment arrangement (Fig. 1) was used in all but the antiphony experiment.

In the antiphony experiment, the test fish's own boatwhistles were used as the playback sounds. Two similar tape recorders were positioned close together (Fig. 2). The left one, connected to the hydrophone and turned on

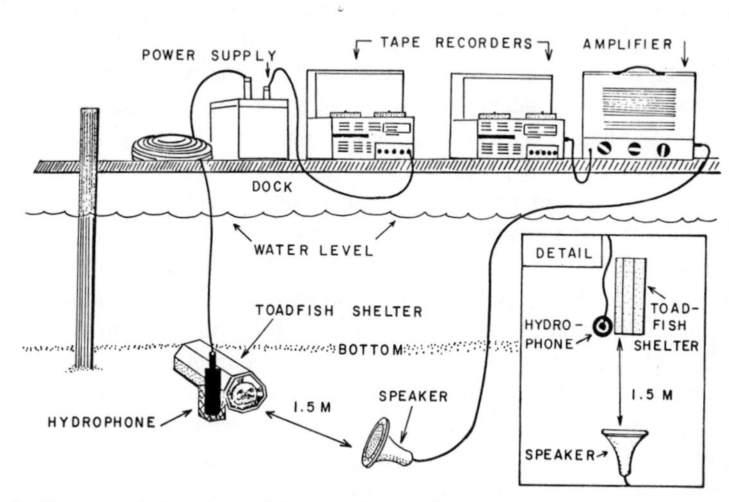

Fig. 1. The normal placement of the sound recording and playback equipment used in the field experiments.

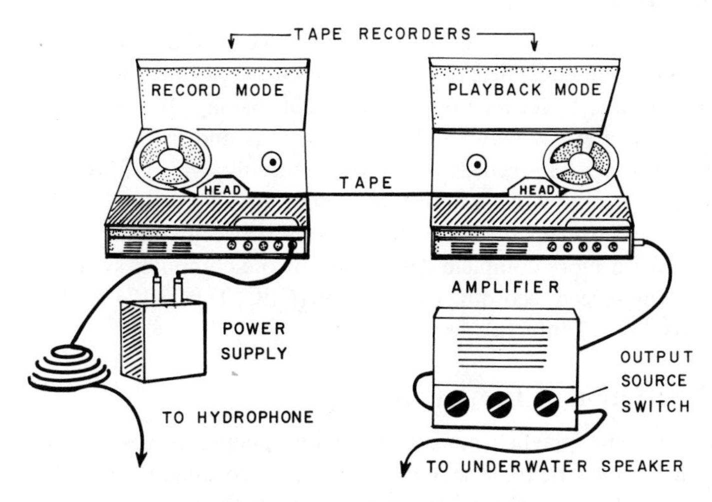

Fig. 2. The special tape recorder arrangement used for playing back a fish's own sounds in the antiphony experiment.

"record," was filled with a new reel of tape; but, instead of winding the tape onto its take-up reel, it was passed over the heads of the right recorder and collected on the take-up reel there. The recorder on the right was turned on playback and connected to the amplifier and underwater speaker, positioned as previously described. All sounds recorded on the left machine were consequently played back by the right machine with a time delay controlled by the tape transport speed and distance separating the machines. However, since only the test fish's calls (and not miscellaneous noises) were to be played back, and each call only once, the playback amplifier was turned on (completing the playback circuit) only at the estimated time when the recorded fish sound (now the playback sound) was about to go over the heads of the right recorder and turned off as soon as the sound passed. The playback sound was then picked up by the hydrophone and recorded on the left machine (but not played back a second time). Thus, the resulting reel of tape on the right recorder contained every sound produced by the test fish followed by a copy of each sound (the playback). Playback gain was adjusted so that there would be an 8 db difference between the original recorded fish call and its recorded duplicate. This made it easy to separate the fish sounds from the playback sounds with an automatic interval analysis technique. The time interval between a fish sound and the following playback was constant for a particular experiment (controlled by tape transport speed and distance between machines), but the interval between a playback sound and the next fish sound depended only on the response time of the fish.

B. Analysis

All the tapes were analyzed by one of two methods. Boatwhistles and grunts were counted by listening to the tapes while playing them back through a B&K Graphic Level Recorder for a visual display. Intervals between fish sounds, and intervals between playback sounds and fish sounds were measured from the B&K records or, more commonly, with a PDP-5 computer. This latter system, with an adjustable amplitude-discriminating circuit, was designed to automatically measure intervals between sounds to the nearest 0.01 sec (for a more complete description of the computer system, see Fish, 1969). A high-speed recording oscillograph (CEC, Type 5-124-A) was used for visual displays of intervals less than 0.1 sec.

C. Description of Playbacks

Boatwhistle playbacks were made by splicing copies of an average-duration boatwhistle (0.38 sec), from a clean recording, together with blank tape to form loops of the desired patterns. Pure-tone patterns were similarly constructed. The tape loops were then re-recorded onto continuous tapes, timed for the desired length playback periods, and preceded by equal lengths of blank tape to provide timed control periods. A description of the various playbacks used in the six different experiments now follows:

1. Calling-Rate Experiment

This experiment utilized two different playback types, subsequently referred to as BW20 and BW24 (Fig. 3). They were boatwhistle patterns with fixed, equal time intervals between all the sounds of each pattern. Both playbacks were presented twice to seven different fish. Playback periods were 5 min long, but they were not preceded by control periods since the objective of this experiment was simply to compare the calling-rate response to two slightly different playback rates. In seven of the 14 replicates, BW20 was

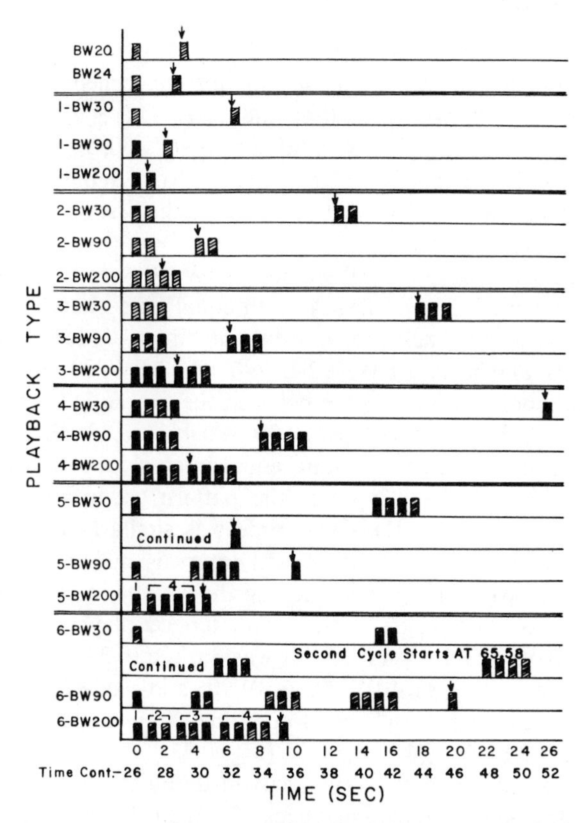

Fig. 3. A schematic of the boatwhistle playback patterns used in the calling-rate and boatwhistle-pattern experiments. The arrows indicate the beginning of the second cycle of the patterns. Use the "time continued" scale for playback types 5-BW30 and 6-BW30.

Table I. Parameters of the Calling-Rate Experiment Playbacks

Playback type	Time interval between BWs[a] (sec)		BW duration (sec)		Cycle time (sec)	No. cycles per 5 min	No. BW per min
BW20	2.62	+	0.38	=	3.0	100	20
BW24	2.12	+	0.38	=	2.5	120	24

[a] BW = Boatwhistle.

presented to the test fish first, followed immediately by BW24, and in the other seven replicates BW20 was first; the order was preselected randomly. Cycle time is defined for this experiment as the time interval from the start of one playback sound to the start of the next. The parameters of the two playbacks are shown in Table I.

2. Boatwhistle-Pattern Experiment

Eighteen playback types were used in this experiment; all were boatwhistle playbacks, differing in pattern and rate. There were six different pattern types and three rates for each (approximately 30, 90, and 200 sounds per 3 min). The 18 playbacks will be referred to as 1-BW30, 1-BW90, 1-BW200; 2-BW30, 2-BW90, 2-BW200, etc., where the first numeral characterizes the pattern type (Fig. 3).

Playback types 1-BW30, 1-BW90, and 1-BW200 were single-boatwhistle patterns (similar to BW20 and BW24) with equal time intervals between all the sounds in each playback type but different intervals for the three playbacks (Table II). Playbacks 2-BW30, 2-BW90, and 2-BW200 were made up of two-boatwhistle bursts with 0.43 sec between the two sounds of each burst; thus, the burst length was 1.19 sec (0.38 + 0.43 + 0.38). Each burst was separated from the next by a constant, equal interval; this interval was different for the three playback types of the pattern group resulting in their different cycle times (Table II). Cycle time was similarly varied in all the following series by making thc interburst intervals different for the three playbacks of each pattern type and keeping the intraburst intervals constant at 0.43 sec. Three-boatwhistle bursts characterized playbacks 3-BW30, 3-BW90, and 3-BW200. The burst length was 2.0 sec (0.38 + 0.43 + 0.38 + 0.43 + 0.38). Pattern types 4-BW30, 4-BW90, and 4-BW200 were similar to the 3-BW series, except each burst contained four boatwhistles, making the burst length 2.81 sec [(4 × 0.38) + (3 × 0.43)].

A single boatwhistle and a four-boatwhistle burst (the same as used in the 4-BW group above) made up playbacks 5-BW30, 5-BW90, and 5-BW200. The interval between the single boatwhistle and the four-boatwhistle burst was the same as between that burst and the next single boatwhistle, starting

Table II. Parameters of the Boatwhistle-Pattern Experiment Playbacks

Playback type	Interburst interval (sec)	Cycle[a] time (sec)	No. cycles per 3 min	No. BW per 3 min	No. BW per min
1-BW30	5.75	6.13	29.4	29	10
1-BW90	1.58	1.96	91.8	92	31
1-BW200	0.55	0.93	193.6	194	65
2-BW30	11.31	12.50	14.4	29	10
2-BW90	2.83	4.02	44.8	90	30
2-BW200	0.55	1.74	103.4	207	69
3-BW30	15.73	17.71	10.2	31	10
3-BW90	4.01	6.01	29.9	90	30
3-BW200	0.68	2.68	67.2	202	67
4-BW30	22.81	25.62	7.0	28	9
4-BW90	5.12	7.93	22.7	91	30
4-BW200	0.67	3.48	51.1	205	68
5-BW30	14.56	32.30	5.6	28	9
5-BW90	3.37	9.91	18.2	91	30
5-BW200	0.61	4.41	40.9	205	68
6-BW30	14.80	65.58	2.7	27	9
6-BW90	3.35	19.78	9.1	91	30
6-BW200	0.64	8.94	20.1	201	67

[a] Cycle time is the sum of all the playback burst lengths and interburst intervals for one complete cycle of the particular pattern. For example, cycle time of the complex pattern 6-BW90 was 0.38 + 3.35 + 1.19 + 3.35 + 2.00 + 3.35 + 2.81 + 3.35 = 19.78 sec.

the second cycle. The most complex patterns were 6-BW30, 6-BW90, and 6-BW200, which combined, in each cycle, a 0.38-sec single boatwhistle, a 1.19-sec two-boatwhistle burst (same as in the 2-BW group), a 2.00-sec three-boatwhistle burst (same as in the 3-BW group), and a 2.81-sec four-boatwhistle burst (same as in the 4-BW group), for a total of ten boatwhistles.

Playback periods were 3 min long, preceded by 3-min preplayback control periods. A run of all 18 patterns, in a randomized order, was played to each of 13 fish.

3. Cycle-Time Experiment

The four playbacks of this experiment were boatwhistle bursts with the 0.38-sec-long boatwhistles being separated by only 0.17 sec within the bursts (Fig. 4). Cycle time from the beginning of one burst to the next was varied for the four playbacks by changing the number of sounds making up the bursts, thus varying burst length, rather than changing the duration of the silent intervals between bursts. The interburst interval was 1.5 sec for all four playbacks. Playback type 2-BW2.4 was a two-boatwhistle burst pattern with a cycle time of about 2.4 sec, type 3-BW3.0 a three-boatwhistle pattern

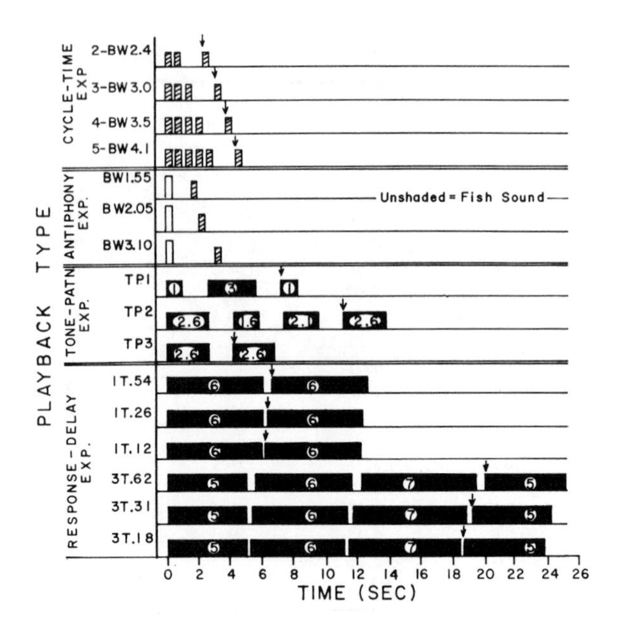

Fig. 4. A schematic of the playbacks used in the cycle-time, antiphony, tone-pattern, and response-time experiments. The arrows indicate the beginning of the second cycle of the playbacks.

Table III. Parameters of the Cycle-Time Experiment Playbacks

Playback type	No. BW in burst	Burst length (sec)	Cycle[a] time (sec)	No. cycles per 5 min	No. BW per min
2-BW2.4	2	0.92	2.42	124	50
3-BW3.0	3	1.48	2.98	101	60
4-BW3.5	4	2.04	3.54	85	68
5-BW4.1	5	2.60	4.10	73	73

[a] Cycle time is the sum of burst length and the constant interburst interval of 1.50 sec.

with a cycle time of about 3.0 sec, etc. (Table III). There were 15 replicates of each pattern (5-min playbacks preceded by 5-min controls). A run consisted of playing all four patterns to the same fish in a randomized order. Eight fish were used for the experiment with two runs consecutively presented to each fish; the second run on one of the eight fish was deleted due to equipment failure.

4. Antiphony Experiment

Instead of prerecorded patterns, the playback sounds for this experiment were the fish's own boatwhistles, each played back to him with small time

Table IV. Parameters of the Tone-Pattern Experiment Playbacks (One Cycle Shown)

Pattern type	TL[a] (sec)	SI[b] (sec)	TL (sec)	SI (sec)	TL (sec)	SI (sec)	Cycle time (sec)
TP1	1.00	1.60	3.00	1.60			7.20
TP2	2.60	1.60	1.60	1.60	2.10	1.60	11.10
TP3	2.60	1.50					4.10

[a] TL = Tone length.
[b] SI = Silent intertone interval.

delays, as previously described. Playback type BW1.55 had a delay time of 1.55 sec; BW2.05 and BW3.10 had delay times of 2.05 and 3.10 sec, respectively (Fig. 4). Twenty-one replicates were made with BW1.55 and BW2.05 (two on each of ten fish, plus one on another fish) and 23 replicates with BW3.10 (two on each of 11 fish, plus one). All of the 3-min playback periods were preceded by 3-min controls. Cycle time from the start of one playback signal to the next for these playbacks was not fixed as it was with all previous playback types, but instead was controlled by the response time of the fish to each playback sound.

5. Tone-Pattern Experiment

In this experiment, the playback patterns were composed of 200-Hz tones instead of boatwhistle bursts. Three different patterns were constructed from the tones (Fig. 4). Each cycle of the first playback type, TP1, contained a 1.0-sec and a 3.0-sec tone separated by equal intertone intervals (Table IV). The second tone pattern, TP2, contained three different-length tones in each cycle, all separated by equal intertone intervals, The third pattern, TP3, was a single tone alternating with a constant intertone interval. Twenty replicates of TP1 were made on ten different fish (two on each) and 16 replicates of TP2 and TP3 on eight of those same ten fish (two on each). Presentation order was randomized. One replicate of TP3 had to be dropped because of background noise interference. Playback periods were 5 min, preceded by 5-min preplayback control periods.

6. Response-Time Experiment

The six 200-Hz tone playback patterns used in this experiment can be divided into two groups (Fig. 4). Playback type 1T.54 of the first group contained one 6-sec tone followed by a 0.54-sec silent interval in each cycle (Table V). Types 1T.26 and 1T.12 of this group similarly contained one 6-sec tone in each cycle, but their silent intervals were 0.26 and 0.12 sec, respectively. The playbacks of the second group contained three different-

Table V. Parameters of the Response-Time Experiment Playbacks (One Cycle Shown)

Playback type	TL[a] (sec)	SI[b] (sec)	TL (sec)	SI (sec)	TL (sec)	SI (sec)
IT.54	6.0	0.54				
IT.26	6.0	0.26				
IT.12	6.0	0.12				
3T.62	5.0	0.62	6.0	0.62	7.0	0.62
3T.31	5.0	0.31	6.0	0.31	7.0	0.31
3T.18	5.0	0.18	6.0	0.18	7.0	0.18

[a] TL = Tone length.
[b] SI = Silent interval duration.

length tones (5.0, 6.0, and 7.0 sec) in each cycle separated by silent intervals equal in duration for a particular playback, but different for each of the three playbacks. The intervals were 0.62, 0.31, and 0.18 sec for playback types 3T.62, 3T.31, and 3T.18, respectively; 3T refers to the three different-duration tones in each cycle (Table V). Twelve replicates were made of each playback in the 1T group on six different fish (two replicates of each type on each fish), and ten replicates of the 3T group on five different fish (two replicates of each type on each fish). One of the 1T replicates could not be used because of noise interference during the experiment. No individual was presented both the 1T and 3T groups, as the latter was not designed until a year after the 1T experiment had been performed and the data analyzed. Playback periods were 5 min long preceded by 5-min controls.

III. RESULTS

A. Calling-Rate Experiment

The Wilcoxon matched-pairs signed-rank test showed that calling rate was significantly greater during playback of BW24 than during the 20%

Table VI. The Number of Boatwhistles Produced During Each of the 14, 5-min Playbacks of the Calling-Rate Experiment[a]

Playback type	Replicate number													
	1	2	3	4	5	6	7	8	9	10	11	12	13	14
BW20	75	36	50	33	52	50	53	62	49	49	27	29	36	24
BW24	80	44	47	42	58	57	58	52	57	62	36	39	42	40

[a] Wilcoxon $T = 12.5$, significant at the 0.05 level.

slower-rate playback BW20. Of the 14 replicate pairs, more boatwhistles were produced during the BW24 playbacks in 12 cases (Table VI).

The data were then pooled for the 14 replicates of each playback type. Intervals were measured between all of the boatwhistles (from the start of one to the start of the next) and grouped according to their durations to the nearest 0.25 sec (Fig. 5). Three large peaks are evident in each histogram; the first resulted from the fish responding once after each playback signal; the second, once after every other signal, and the third, once after every third signal. The first and highest peak of the BW20 histogram falls in the 3.00–3.24 sec interval. Cycle time of the playback was 3.0 sec (Table I), indicating that many times the test fish responded after each playback signal; the second peak at 5.50–5.74 sec is nearly twice the playback cycle time, and the third peak at about 9 sec almost thrice. Histogram BW24 (Fig. 5) shows that the test fish did not respond to every cycle of this playback (cycle time 2.5 sec) as often as when the cycle time was 3.0 sec. Instead, most responses came after every other cycle, causing a very high second peak at 4.50–4.74 sec, almost twice the cycle time of the BW24 playback. The first

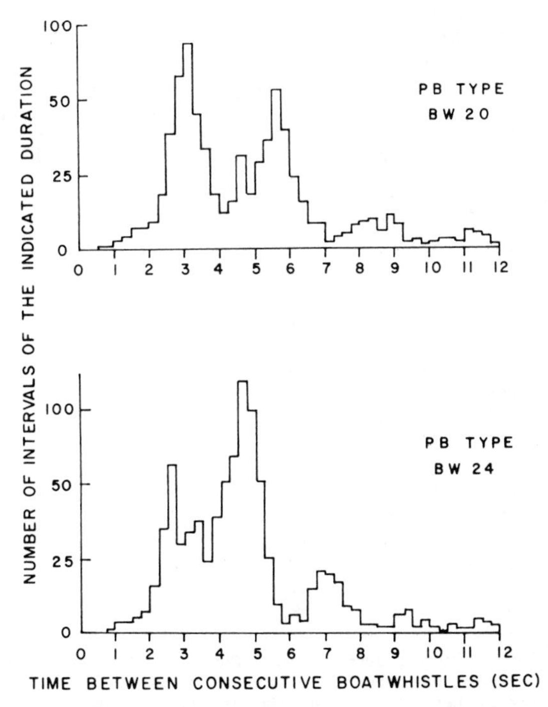

Fig. 5. Histograms showing the distribution of intervals between all the boatwhistles produced during playbacks BW20 and BW24 of the cycle-time experiment. Class interval size is 0.25 sec.

and third peaks are much smaller than the second. In both histograms, the second peaks are 0.25 sec short of two playback cycles; when responses occurred only after every other playback signal, they began slightly before the expected time. There was no significant difference in the number of grunt-type sounds produced with the two playback types.

B. Boatwhistle-Pattern Experiment

The Wilcoxon matched-pairs signed-rank test was used to compare the number of boatwhistles produced during the control and playback periods of the 13 replicates of each of the 18 playback types. There was no significant change in calling with any of the 30 BW per 3 min playbacks regardless of pattern type (Table VII). Calling rate increased from control to playback with playbacks 1-BW90, 2-BW90, 4-BW90, and 6-BW90 of the 90 BW per 3 min series, and with all six of the fastest-rate, 200 BW per 3 min, playbacks.

Table VII. Results of Boatwhistle-Pattern Experiment[a]

Playback type	Total BW all reps.		Median BW per rep.		No. reps.			Wilcoxon T
	Con	PB	Con	PB	Increase	Decrease	No change	
1-BW30	327	319	30	30	6	6	1	37.0
2-BW30	438	417	34	34	3	9	1	22.5
3-BW30	347	367	28	30	8	5	—	35.5
4-BW30	396	401	31	34	7	5	1	36.0
5-BW30	360	365	32	33	4	9	—	35.0
6-BW30	392	357	37	33	5	8	—	24.5
1-BW90	462	591	34	46	11	2	—	6.5[b]
2-BW90	384	466	33	35	9	4	—	19.0[b]
3-BW90	371	455	30	34	8	4	1	18.0
4-BW90	362	489	30	36	8	4	1	11.0[b]
5-BW90	354	425	32	33	9	4	—	26.5
6-BW90	371	485	32	37	10	3	—	10.5[b]
1-BW200	283	453	24	38	9	1	3	4.0[b]
2-BW200	366	507	34	38	10	3	—	12.0[b]
3-BW200	337	468	29	36	10	3	—	9.0[b]
4-BW200	347	531	29	38	10	2	1	6.5[b]
5-BW200	389	486	32	35	9	3	1	10.0[b]
6-BW200	407	517	35	39	9	3	1	12.0[b]

[a] Total number of boatwhistles produced during all the control (Con) and playback (PB) periods of the boatwhistle-pattern experiment; the number of replicates that showed an increase, decrease, or no change in number of boatwhistles from Con to PB; and the Wilcoxon T values for the changes in boatwhistle calling. For the number of sounds produced during each replicate, see Fish (1969).

[b] Significant increase in BW from Con to PB (0.05 level).

In addition to comparing the total number of boatwhistles produced during the control and playback periods, a sequential half-minute analysis was performed. During the field experiments, it appeared that the 200 BW per 3 min playbacks actually reduced calling for a short time at the start of the playback period, even though all six patterns caused a significant increase in calling when the entire control and playback periods were compared. When the last minute of preplayback was compared to the first minute of playback, a significant decrease in calling was noted with playbacks 5-BW200 and 6-BW200 (Wilcoxon test). However, when only the last 30 sec of the control periods was compared to the first 30 sec of playback, a significant decrease occurred with all six of the 200 BW per 3 min playbacks. None of the 30 or 90 BW per 3 min playbacks caused a significant decrease during either the first 30 sec or first minute of playback.

Increased calling during playback of the 90 and 200 BW per 3 min patterns did not simply result from shorterning all the intervals between calls by a constant amount. Rather, calling patterns were established that were closely related to the cycle times and patterns of the playbacks presented to the fish. Figures 6 through 11 show the distributions of intervals between boatwhistles for the pooled 13 replicates of each playback type. Each figure contains three sets of curves representing the three playback rates of the particular pattern. The left sections of the figures correspond to the 30 BW per 3 min rate, the center sections to the 90 BW per 3 min rate, and the right sections to the 200 BW per 3 min rate. The lower curves show the difference between the playback and control period histograms; each interval of the control period histogram was subtracted from the corresponding interval of the playback histogram.

Refer to Fig. 3 and Table II (cycle times) during the following discussion of Figs. 6–11. It can be seen from the left sections of the six figures that the 30 BW per 3 min playbacks caused very little change in calling pattern. The difference curves oscillate closely around zero.

Except for 5-BW90, the 90 BW per 3 min playbacks and 200 BW per 3 min playbacks all caused a significant change in calling pattern, as shown by the changes in histogram shape from control to playback and the numerous high peaks in the difference curves. The location of the peaks corresponds to the predicted location based on the cycle times of the playbacks.

For example, the high peak in 1-BW90 (Fig. 6, center) resulted from the test fish responding frequently after every other playback signal; the cycle time of the playback was 1.96 sec (Table II); the highest peak of the histogram and difference curve is in the interval 3.50–3.99 sec (two times the playback cycle time). A smaller peak at 2.00–2.49 sec resulted from occasional responses after every playback signal, and the third peak at 5.50–5.99 sec (three times the cycle time) from responses after every third signal. The three peaks

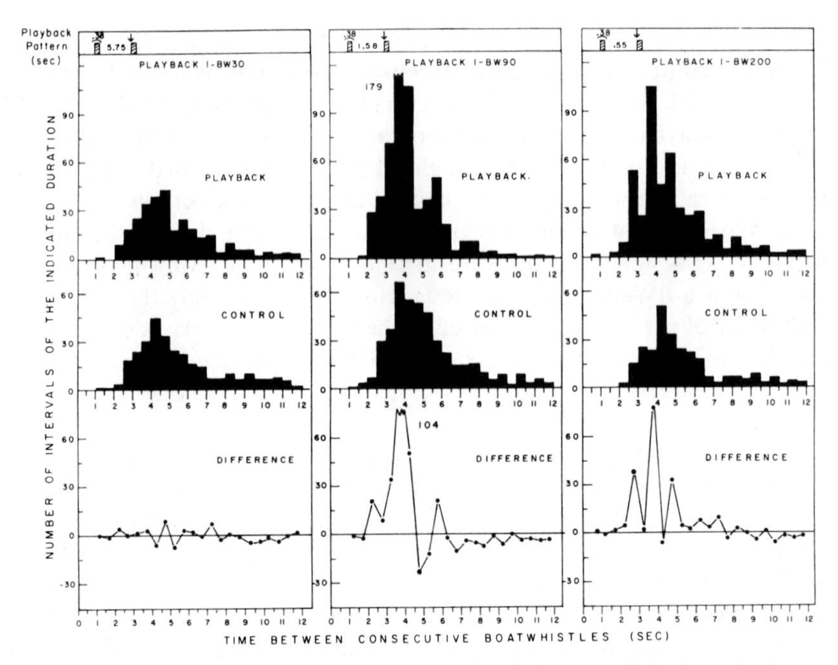

Fig. 6. Histograms showing the distribution of intervals between all the boatwhistles produced during the control and playback periods of pattern types 1-BW30, 1-BW90, and 1-BW200. The lower curves represent the difference between the playback and control period histograms for very interval (class interval size is 0.50 sec). Schematics of the playbacks above each set of figures (not to scale) show the playback burst durations and interburst intervals in seconds for the three different playback types. Arrows indicate the beginning of the second cycle of playback.

in the playback histogram and difference curve for 1-BW200 (Fig. 6. right) occur at 2.50–2.99, 3.50–3.99, and 4.50–4.99 sec, corresponding to three, four, and five times the very short (0.93 sec) cycle time, respectively. Generally, the fish responded after every fourth signal (second peak), but also quite often after every third or every fifth.

The peak occupying the two intervals 3.50–3.99 and 4.00–4.49 sec in Fig. 7 for playback 2-BW90 is centered at 4 sec, equal to the cycle time (4.02 sec) of the playback, indicating that the fish responded most often after each two-boatwhistle playback burst. The highest peak for 2-BW200 is located at 3.00–3.49 sec, twice the cycle time (1.74 sec) of the playback; most responses occurred after every other playback signal.

The peak at 5.50–5.99 sec in the histogram for playback 3-BW90 (Fig. 8) is considerably displaced to the right of the control-period histogram peak. It corresponds to one long cycle (6.01 sec) of the three-boatwhistle-burst playback. The high occurrence of intervals just preceding the peak resulted from the test fish not waiting the full 6 sec to respond; instead, they called earlier in the interburst intervals. The two peaks of the 3-BW200 playback period histogram at 2.50–2.99 and 5.00–5.49 sec correspond to one and two times the playback cycle time (2.68 sec), respectively. Generally, a fish would respond after every other playback burst during the early part of the playback period, then increase its calling rate for about 1 min, responding after each burst, and finally return to the slower calling rate for the remainder of the playback period.

The peaks in 4-BW90 (Fig. 9) do not correspond to the long cycle time of 7.93 sec. The first peak (2.50–2.99 sec) resulted from the fish placing two calls, 2.50–2.99 sec apart, in the 5.12-sec interburst intervals, and the second peak from placing single calls in the succeeding intervals. The highest peak

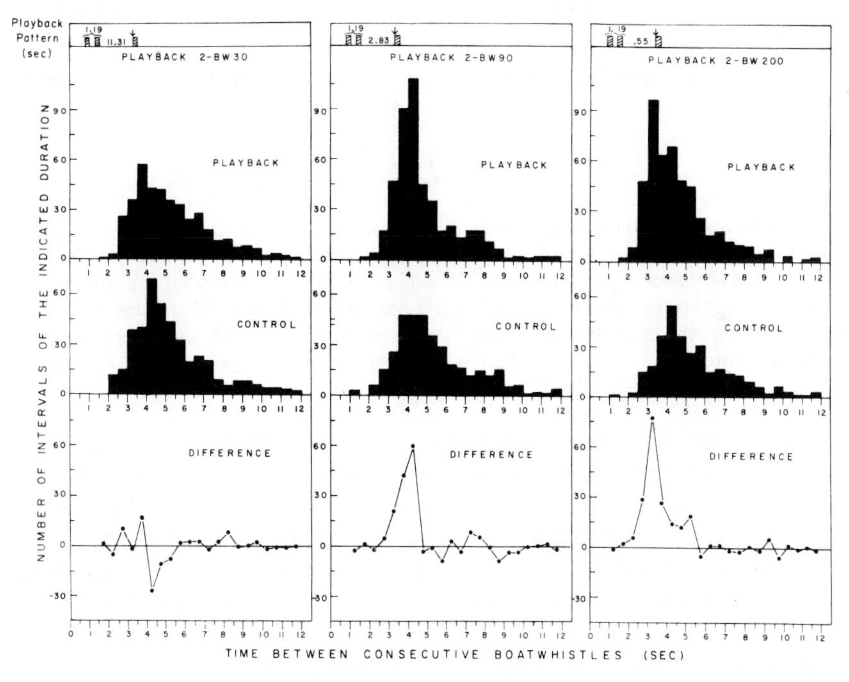

Fig. 7. Histograms showing the distribution of intervals between all the boatwhistles produced during the control and playback periods of pattern types 2-BW30, 2-BW90, and 2-BW200. The lower curves represent the difference between the playback and control period histograms for every interval (class interval size is 0.50 sec). Schematics of the playbacks above each set of figures (not to scale) show the playback burst durations and interburst intervals in seconds for the three different playback types. Arrows indicate the beginning of the second cycle of playback.

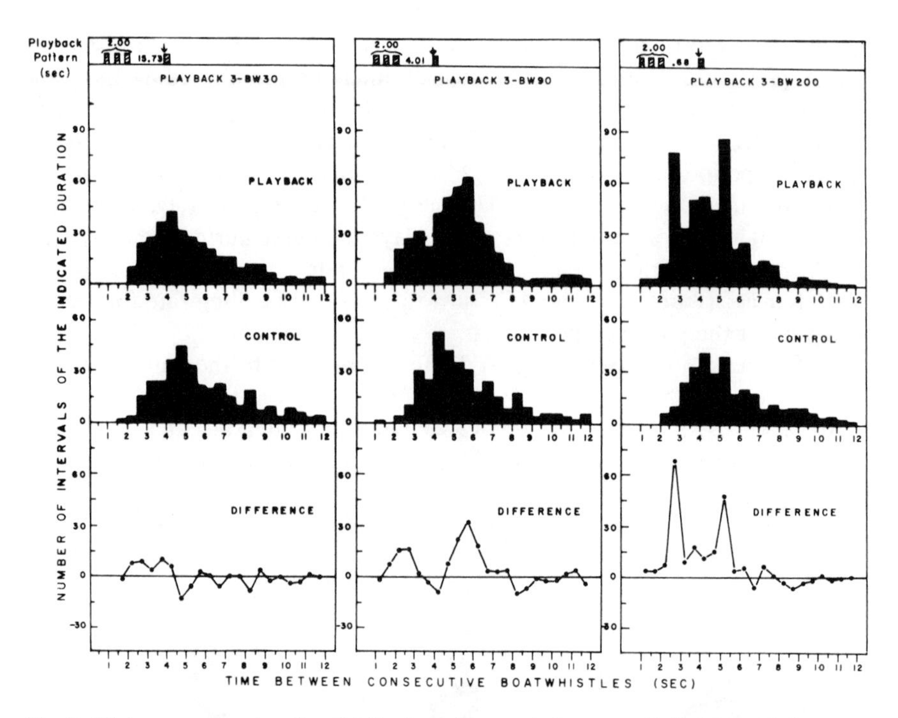

Fig. 8. Histograms showing the distribution of intervals between all the boatwhistles produced during the control and playback periods of pattern types 3-BW30, 3-BW90, and 3-BW200. The lower curves represent the difference between the playback and control period histograms for every interval (class interval size is 0.50 sec). Schematics of the playbacks above each set of figures (not to scale) show the playback burst durations and interburst intervals in seconds for the three different playback types. Arrows indicate the beginning of the second cycle of playback.

in any of the 18 difference curves occurred with 4-BW200; the single peak falling in the interval 3.00–3.45 sec corresponds to the playback cycle time of 3.48 sec. No high secondary peaks are present, indicating that the fish nearly always responded after each four-boatwhistle playback burst.

Playbacks 5-BW30, 5-BW90, and 5-BW200 had a single boatwhistle and a four-boatwhistle burst in each cycle. As previously noted, 5-BW90 did not cause a change in calling pattern (Fig. 10). However, there is an outstanding difference in the control and playback period histograms for 5-BW200. A schematic of three cycles of playback 5-BW200 (Fig. 12a) shows where the responses fell (resulting in the histogram peaks) in relation to the playback pattern. The second and highest peak (4.00–4.49 sec) in the histogram was due to a response falling at point A (Fig. 12a) and the next at C or a response at B and the next at D (both equal to one cycle). The first peak in the histogram (3.00–3.49 sec) corresponds to a response at A followed by one at B (2.81 + 0.61 = 3.42 sec), and the last peak (5.00–5.49 sec) a response at B and the next at E (4.41 + 0.38 + 0.61 = 5.40 sec). A response at B was never followed by one at C.

The 6-BW playback series (Fig. 11), composed of a single boatwhistle, a two-boatwhistle burst, a three-boatwhistle burst, and a four-boatwhistle burst, was too complex to elicit a change in calling pattern, even though both 6-BW90 and 6-BW200 caused an increase in calling rate. The playback period histograms are higher (due to the increase in calling rate), but have the same general shape as the control period histograms. The single, very wide peak in the histogram resulted from responses being distributed equally in the silent interburst intervals of the playbacks. A schematic of two cycles of 6-BW200 shows some of the many possible interboatwhistle intervals that could have existed without responses overlapping the playback signal bursts (Fig. 12b).

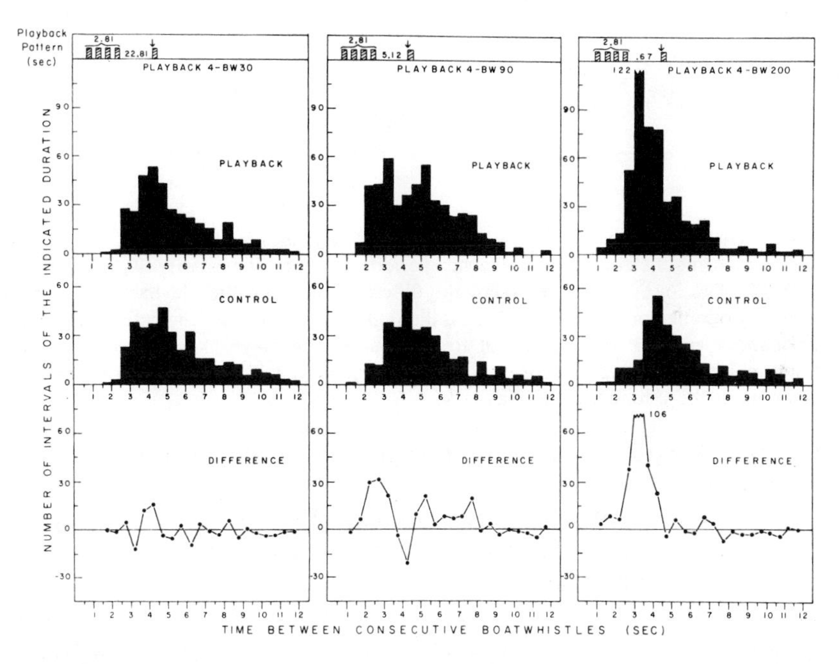

Fig. 9. Histograms showing the distribution of intervals between all the boatwhistles produced during the control and playback periods of pattern types 4-BW30, 4-BW90, and 4-BW200. The lower curves represent the difference between the playback and control period histograms for every interval (class interval size is 0.50 sec). Schematics of the playbacks above each set of figures (not to scale) show the playback burst durations and interburst intervals in seconds for the three different playback types. Arrows indicate the beginning of the second cycle of playback.

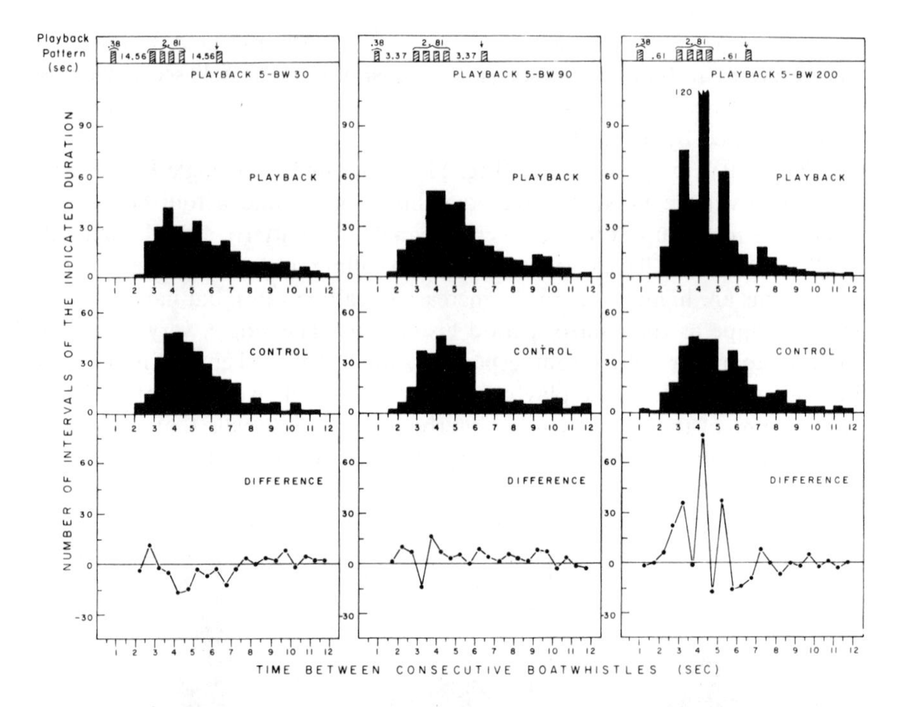

Fig. 10. Histograms showing the distribution of intervals between all the boatwhistles produced during the control and playback periods of pattern types 5-BW30, 5-BW90, and 5-BW200. The lower curves represent the difference between the playback and control period histograms for every interval (class interval size is 0.50 sec). Schematics of the playbacks above each set of figures (not to scale) show the playback burst durations and interburst intervals in seconds for the three different playback types. Arrows indicate the beginning of the second cycle of playback.

As playback pattern complexity increased, so did the number of possible interboatwhistle intervals.

The specific location of every boatwhistle produced during the playback periods of each of the 18 playback types was determined from the B&K graphic output. Any response not falling within an interburst interval was counted as an overlap (Fig. 12c). As may be expected, the number of overlaps increased as the amount of time available to respond without overlapping playback sounds decreased.

In addition to counting the ovelapping boatwhistles, the specific location of all nonoverlapping responses (i.e., those falling in the interburst intervals) was noted for the complex 5-BW and 6-BW playbacks. Approximately half of the responses fell after the single boatwhistle and half after the four-boatwhistle burst for playback types 5-BW30 and 5-BW90. However, when playback type 5-BW200 was presented to the test fish, they responded 71%

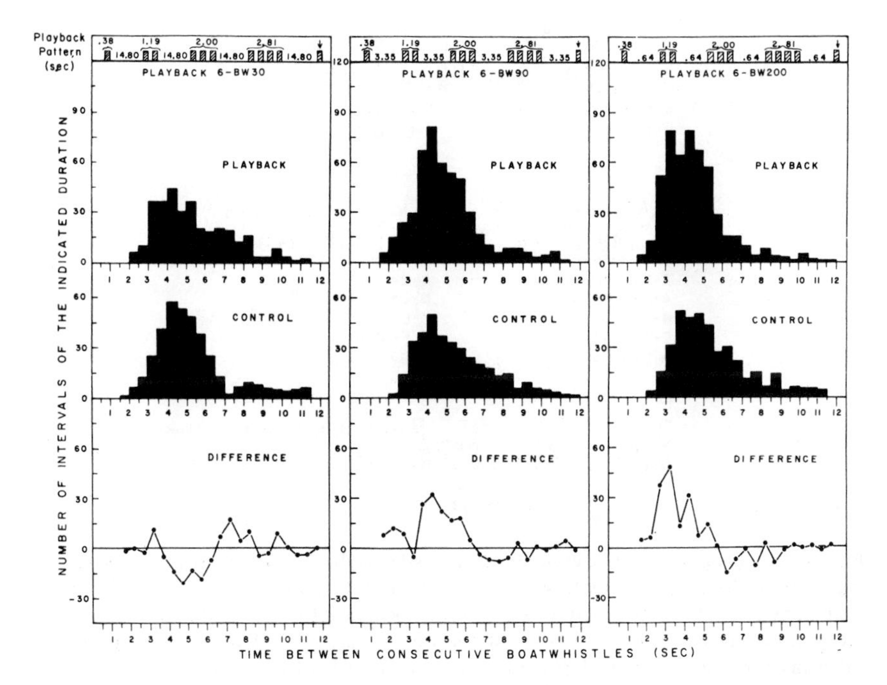

Fig. 11. Histograms showing the distribution of intervals between all the boatwhistles produced during the control and playback periods of pattern types 6-BW30, 6-BW90, and 6-BW200. The lower curves represent the difference between the playback and control period histograms for every interval (class interval size is 0.50 sec). Schematics of the playbacks above each set of figures (not to scale) show the playback burst durations and interburst intervals in seconds for the three different playback types. Arrows indicate the beginning of the second cycle of playback.

of the time after the four-boatwhistle burst and only 29% of the time after the single boatwhistle, even though the interburst intervals were equal. This distribution is significantly different from the expected equal distribution after the single- and four-boatwhistles bursts ($\chi^2 = 23.4$; significant at the 0.05 level, two-tailed test).

Approximately 25% of the total nonoverlapping responses fell after each of the four different-length bursts of the most complex patterns 6-BW30, 6-BW90, and 6-BW200. That is, the fish made no attempt to imitate the pattern or even space their responses in a specific location with respect to the different-length bursts as they did with playback type 5-BW200.

C. Cycle-Time Experiment

Playback types 2-BW2.4 (two-boatwhistle burst with a 2.42-sec cycle time), 3-BW3.0 (three-boatwhistle burst with a 2.98-sec cycle time), 4-BW3.5

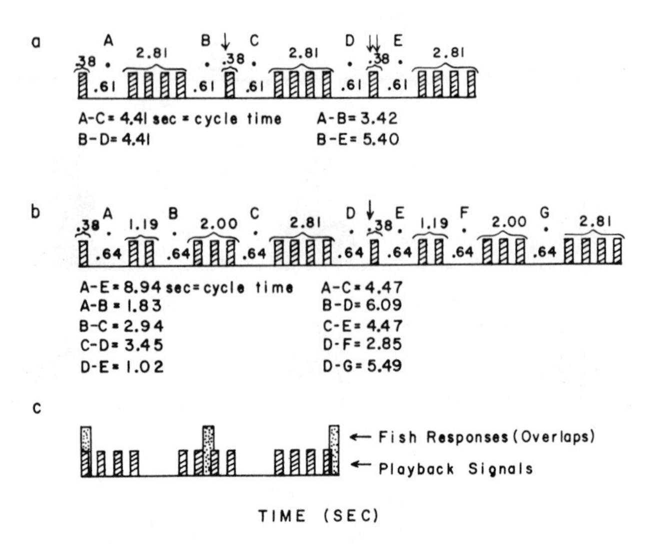

Fig. 12. **(a)** Diagram of three cycles of boatwhistle playback 5-BW200 with letters indicating common locations of boatwhistle responses, resulting in the peaks in the 5-BW200 histogram (Fig. 10). Single and double arrows are located at the start of the second and third cycles of the playbacks. **(b)** Two cycles of boatwhistle playback 6-BW200 showing some of the many possible interboatwhistle intervals (time between letters). **(c)** Diagram of some response locations that are defined as overlaps.

(four-boatwhistle burst with a 3.54-sec cycle time), and 5-BW4.1 (five-boatwhistle burst with a 4.10-sec cycle time) all had the same silent interburst interval duration of 1.50 sec. Each playback caused a significant increase in boatwhistle calling from the preplayback controls to the playback periods (Table VIII).

The effect of each playback type on the calling pattern of the test fish

Table VIII. Results of Cycle-Time Experiment[a]

Playback type	Total BW all reps.		Median BW per rep.		No. Reps.			Wilcoxon T
	Con	PB	Con	PB	Increase	Decrease	No change	
2-BW2.4	366	781	20	51	14	—	1	0[b]
3-BW3.0	434	733	31	52	14	1	—	5.0[b]
4-BW3.5	465	816	33	59	13	1	1	3.0[b]
5-BW4.1	477	786	29	58	13	2	—	3.0[b]

[a] Total number of boatwhistles produced during all the control (Con) and playback (PB) periods of the cycle-time experiment; the number of replicates that showed an increase, decrease, or no change in number of boatwhistles from Con to PB; and the Wilcoxon T values for the changes in boatwhistle calling. For the number of sounds produced during each replicate, see Fish (1969).

[b] Significant increase in BW from Con to PB (0.05 level).

can be seen in Fig. 13. The histograms show the distributions of time intervals between fish sounds for the pooled 15 replicates of each playback type. While the histograms for the control periods of the four playback types are very similar, those for the playback periods have very different shapes; the peaks correspond directly to the playback cycle times. The first peak of the 2-BW2.4 playback histogram is in the interval 2.50–2.74 sec, which is only slightly longer than the playback cycle time of 2.42 sec; the second and highest peak at 4.50–4.74 sec is slightly shorter than two cycles; and the third peak at 7.00–7.24 sec is equivalent to three cycles. The fact that the second peak is larger than the first indicates that the test fish called more often after every other playback cycle than after each cycle; cycle time was too short to permit continuous, equal-interval calling.

In histogram 3-BW3.0, the first peak is in the interval 2.75–2.99 sec, equivalent to one playback cycle (cycle time 2.98 sec); the second peak in the interval 5.75–5.99 corresponds to two cycles. The first peak is larger

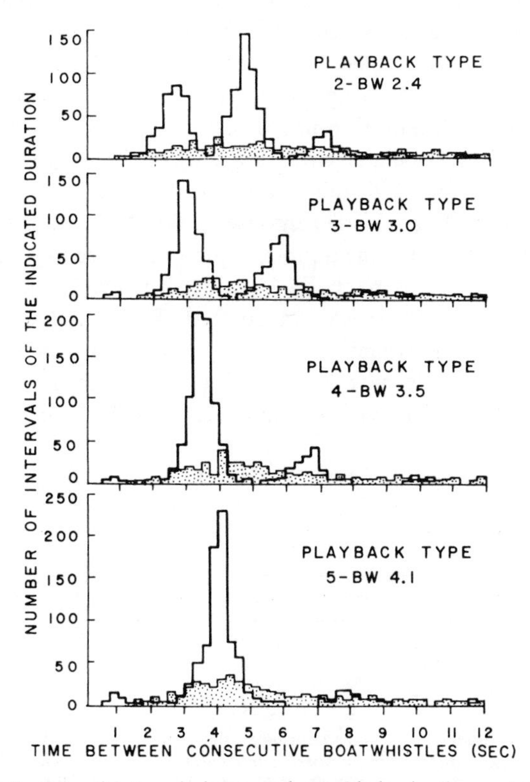

Fig. 13. The distribution of intervals between boatwhistles in the control and playback periods for the four cycle-time experiment playback types. Class interval size is 0.25 sec. Speckled curves are for the control periods, plain curves for the playbacks.

than the second because responses occurred more often after every playback burst than after every other one. The third peak has now disappeared. The ratio of the first peak to the second becomes even greater in the 4-BW3.5 histogram. The first peak occurs in the interval 3.25–3.49 sec (cycle time 3.54 sec); the second, at two cycles, has now become quite small. Cycle time of playback type 5-BW4.1 was optimum (4.10 sec) for a response after each playback burst for several successive cycles. The first peak at 4.00–4.24 sec is much larger than the second peak.

D. Antiphony Experiment

A significant increase in boatwhistle calling from control to playback occurred with all three playback delay times: 1.55, 2.05, and 3.10 sec (Table IX).

To determine if the change in calling rate from control to playback became greater as the playback signal delay time was decreased from 3.10 to 1.55 sec, the number of boatwhistles produced during the control period of each replicate was subtracted from the number produced during the corresponding playback. The resulting differences were then compared for the two delay times with the Mann–Whitney U test. The computed Z (N larger than 20) was 2.35, and its associated probability was 0.009; the increase in calling was significantly greater with the shorter playback delay.

Of the total 851 boatwhistle responses produced during the 21 replicate playbacks of BW1.55, only one double response occurred—a double response being defined as a second boatwhistle emitted before the delayed playback of the preceding boatwhistle was transmitted. This is not surprising in view of the short 1.55-sec playback signal delay time. For a double response to occur, the test fish would have had to call twice within only 1.55 sec. In the 21 replicates of BW2.05 (playback signal delay time 2.05 sec), there were seven

Table IX. Results of Antiphony Experiment[a]

Playback type	Total BW all reps.		Median BW per rep.		No. reps.			Wilcoxon T
	Con	PB	Con	PB	Increase	Decrease	No change	
BW1.55	609	851	31	42	21	—	—	0[b]
BW2.05	510	747	23	37	19	2	—	12.5[b]
BW3.10	702	864	32	37	20	3	—	20.0[b]

[a] Total number of boatwhistles produced during all the control (Con) and playback (PB) periods of the antiphony experiment; the number of replicates that showed an increase, decrease, or no change in number of boatwhistles from Con to PB; and the Wilcoxon T values for the changes in boatwhistle calling. For the number of sounds produced during each replicate, see Fish (1969).

[b] Significant increase in BW from Con to PB (0.05 level).

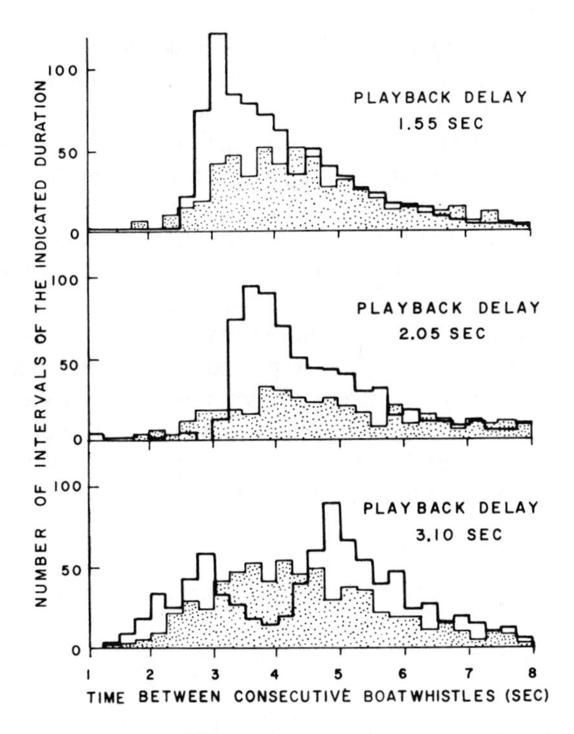

Fig. 14. The distribution of intervals between boatwhistles in the control and playback periods for the three different playback delays used in the antiphony experiment. Class interval size is 0.25 sec. Speckled curves are for the control periods, plain curves for the playbacks.

double responses out of the total 747 boatwhistles produced. However, in the 23 replicates of BW3.10 (playback signal delay time 3.10 sec), double responses occurred 168 times (336 boatwhistles out of a total of 864). Rather than waiting 3.10 sec or longer and responding after the next delayed playback signal, the fish often produced a second sound within the 3.10 sec.

The distributions of intervals between calls were essentially the same for the control periods of the three playbacks (Fig. 14). During the playback periods, however, intervals between boatwhistles were longer with the longer playback delay times. The difference was actually equal to the difference in the delay times of the three playbacks. The peak of the BW2.05 histogram (Fig. 14) occurs 0.50 sec later than the peak of BW1.55; playback signal delay time was 0.50 sec longer for BW2.05 than BW1.55. The higher mode of BW3.10 occurs 1.25 sec later than the peak of BW2.05, only 0.20 sec longer than would be expected from the difference in playback delay times (1.05 sec). The first peak of BW3.10 resulted from the short intervals between the 168 double responses.

In order for the shifts of the histogram peaks in Fig. 14 to be equal to the difference in playback signal delay times, the distributions of intervals between the playback signals and the following boatwhistle responses must have been about the same for all three delay times. Figure 15 shows this to be true except for a slight 0.25-sec shift to the right in the BW3.10 histogram. This accounts for the greater than expected shift to the right in the peak of the interboatwhistle interval histogram for BW 3.10 in Fig. 14.

The shift to the right in the interboatwhistle interval histogram modes with increased playback signal delay time was due then only to the increase in playback delay and not to a difference in the response time of the fish. The constant response time of 1.5–2.0 sec phased well with the playback signal when the playback delay was 1.55 or 2.05 sec, resulting in frequent alternation for several cycles. However, the precision was not maintained for long, as evidenced by the wide-peaked histograms; compare them with the narrow peaks in Fig. 13 for fixed-interval playbacks. When the playback delay was

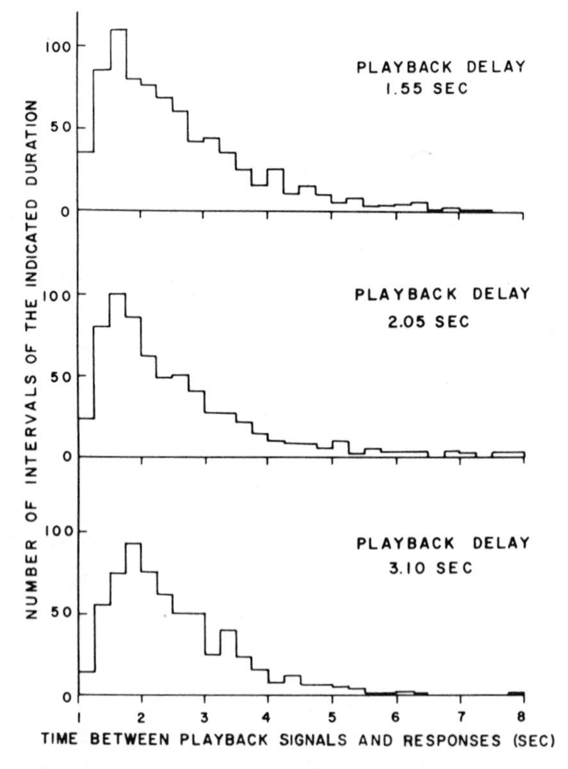

Fig. 15. The distribution of intervals from the beginning of the playback signals to the beginning of the following boatwhistle responses for the three different playback delays used in the antiphony experiment. Class interval size is 0.25 sec.

extended to 3.10 sec, alternation occurred less frequently and the response time of 1.5–2.0 sec allowed enough time for a second response to be emitted (on 168 occasions) before the delayed playback of the first sound was transmitted.

E. Tone-Pattern Experiment

Each of the three 200-Hz tone playbacks (TP1, TP2, and TP3) caused a significant increase in boatwhistle calling from control to playback (Table X).

The tone pattern TP1 (Fig. 4 and Table IV) was similar to the boatwhistle-burst playback 5-BW200 (Fig. 3 and Table II). One cycle of both playbacks consisted of a short and a long playback burst (or tone) separated by an interburst (or intertone) interval which was the same between the short and long burst (or tone) as between the long and the next short. Of the total 805 boatwhistles produced during the playback periods of the 20 replicates of TP1, 759 fell in the intertone intervals, leaving only 46 overlaps. Responses were not distributed equally after the short and long tones (even though the intertone intervals were equal); 65% fell after the 3-sec tones and 35% after the 1-sec tones. This distribution is similar to that for playback 5-BW200, where 71% fell after the four-boatwhistle bursts and 29% after the single boatwhistle.

The distributions of intervals between boatwhistles for the control and playback periods of the three tone patterns are shown in the histograms of Fig. 16. As with the boatwhistle-burst playbacks, the peaks can be explained in terms of the playback patterns. The second and smaller peak of TP1, in the interval 6.75–6.99 sec, corresponds to one cycle of the playback (cycle time 7.2 sec). This peak resulted from the fish consecutively responding after two short tones, skipping the long; or consecutively responding after

Table X. Results of Tone-Pattern Experiment[a]

Playback type	Total BW all reps.		Median BW per rep.		No. Reps.			Wilcoxon T
	Con	PB	Con	PB	Increase	Decrease	No change	
TP1	427	805	19	43	20	—	—	0[b]
TP2	395	635	27	39	14	2	—	15.5[b]
TP3	271	533	14	45	14	—	1	0[b]

[a] Total number of boatwhistles produced during all the control (Con) and playback (PB) periods of the tone-pattern experiment; the number of replicates that showed an increase, decrease, or no change in number of boatwhistles from Con to PB; and the Wilcoxon T values for the changes in boatwhistle calling. For the number of sounds produced during each replicate, see Fish (1969).

[b] Significant increase in BW from Con to PB (0.05 level).

two long tones, skipping the short. In many cases, however, a response followed each playback tone, accounting for the first and largest peak of the histogram. The center of this peak at about 3.5 sec resulted from the most common placement of responses within the intertone intervals, as shown by the schematic in Fig. 17a. Most of the responses following the long tones came soon (0.2–0.5 sec) after the tones ended, but those following the short tones lagged by more than a second. The wide peak of the TP1 histogram was caused by some variation in response location within the intertone intervals. For example, if response B in the schematic had fallen 0.2 sec later, the two resulting interboatwhistle intervals would have been 3.9 and 3.3 sec.

The playback-period histogram for TP2 (Fig. 16) also has a wide peak, resulting from the complex playback pattern. As pattern complexity increased, the number of possible interboatwhistle intervals that could exist, without the responses overlapping the playback signals, also increased. A schematic of TP2 with some possible response locations demonstrates this (Fig. 17b). Even the few interboatwhistle intervals shown would be enough

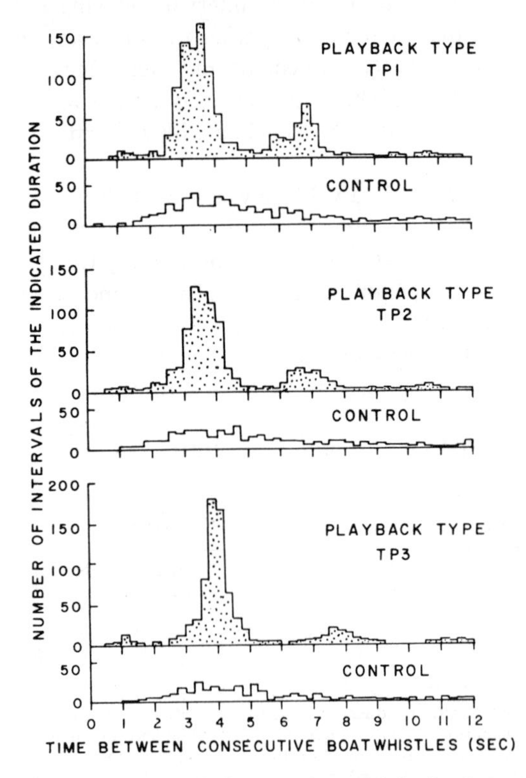

Fig. 16. The distribution of intervals between boatwhistles in the control and playback periods for the three tone-pattern experiment playback types. Class interval size is 0.25 sec.

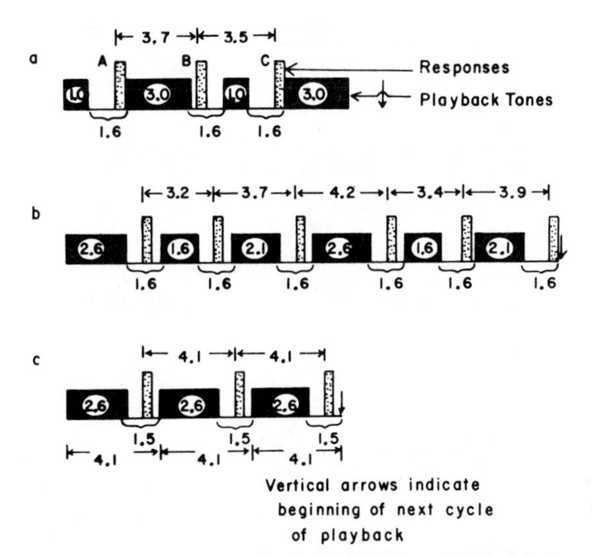

Fig. 17. (a) Schematic of two cycles of tone-pattern playback TP1 with the most common response locations. All tone lengths and intervals between responses are in seconds. **(b)** Schematic of two cycles of tone-pattern playback TP2 with some of the possible response locations. **(c)** Schematic of three cycles of tone-pattern TP3 showing the rhythmic response pattern. Vertical arrows indicate beginning of next cycle.

to account for the wide histogram peak of 3.00–4.24 sec. Similarly, wide peaks were evident for the complex boatwhistle-burst patterns such as the 6-BW series (Fig. 11).

Playback type TP3 was designed with the same cycle time (4.10 sec) as the five-boatwhistle-burst playback 5-BW4.1 (Table III) of the cycle-time experiment. Also, the tone length of TP3 (2.60 sec) was equal to the burst length of 5-BW4.1, and consequently the intertone intervals of the former equalled the interburst intervals of the latter. The narrow peak of the histogram for TP3 (Fig. 16) centered at 4.00 sec corresponds to the playback cycle time of 4.10 sec (Fig. 17c) and is very similar in shape and location to the peak of the histogram for 5-BW4.1 (Fig. 13).

Single narrow peaks were characteristic of all simple playbacks (each cycle containing only one playback burst or tone) which had optimum cycle times (3.5–4.5 sec). Also, tone or burst durations greater than 2 sec increased the precision of the response rhythm, thus reducing the width of the histogram peaks, by preventing calling during a large portion of a playback cycle. Such single peaks were observed for playbacks 2-BW90 (Fig. 7), 4-BW200 (Fig. 9), 4-BW3.5 and 5-BW4.1 (Fig. 13), and TP3 (Fig, 16). All of these were simple patterns with cycle times of 3.5–4.5 sec, and all except 2-BW90 had burst lengths of 2 sec or greater. It made no difference whether the playback

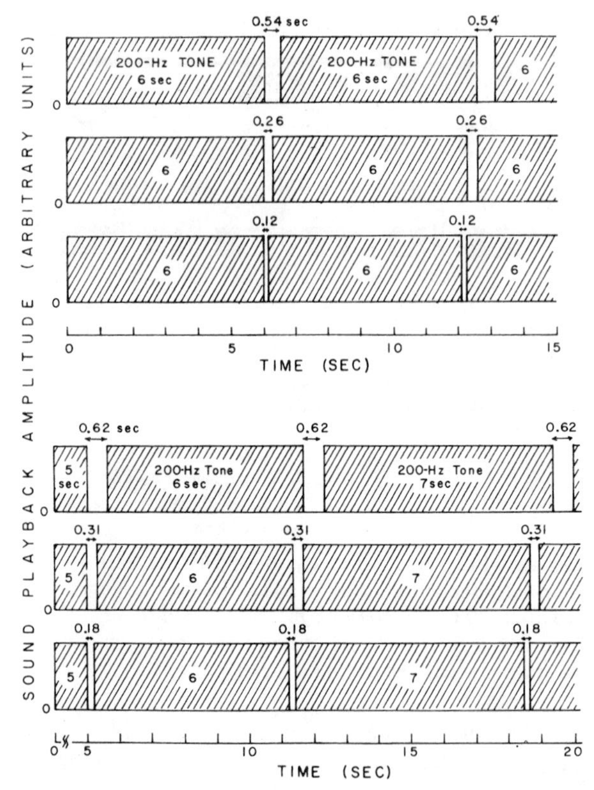

Fig. 18. Two cycles of each of the three 6-sec tone playbacks (1T.54, 1T.26, 1T.12) of the response-time experiment and one cycle of each of the 5-, 6-, and 7-sec tone playbacks (3T.62, 3T.31, 3T.18).

material was a boatwhistle burst or a pure tone. All significantly increased calling from control to playback and established a precise response rhythm.

The number of boatwhistles produced during the control period of each replicate of the tone pattern TP3 (2.60-sec tone) was substracted from the number during the playback. These 15 differences were compared using the Mann–Whitney U test to the 15 differences obtained in the same way for the 2.60-sec five-boatwhistle-burst playback (5-BW4.1) replicates. While, as previously demonstrated, both playback types significantly increased calling over the preplayback controls, there was no significant difference (0.05 level, two-tailed test) in the degree of increase resulting from the playback types (the computed U was 100). That is, the 2.60-sec pure tones, separated by silent intervals of 1.5 sec, had the same stimulatory effect as the 2.60-sec five-boatwhistle bursts, separated by 1.5 sec.

F. Response-Time Experiment

Each cycle of playback 1T.54 consisted of a 6-sec tone followed by a 0.54-sec silent period (Fig. 18). Of the total 286 boatwhistles produced during the 11 playback periods of 1T.54, 245 (86%) actually started in the silent periods. The remaining 14% began while the tones were playing. Most boatwhistles were 0.25–0.40 sec long; thus, it was possible for them to fit entirely in the silent intervals if they started shortly after the tones ended (Fig. 19a). When a response began later in the gap, it overlapped the start of the next tone (Fig. 19b). While few boatwhistles could fit entirely in the 0.26-sec silent periods of playback 1T.26, 212 of the 228 calls produced (93%) started in the gaps (Fig. 19c). When the null period was reduced to 0.12 sec, only 128 boatwhistles were produced during the 11 replicates; however, 56 (43%) still began in the small silent intervals.

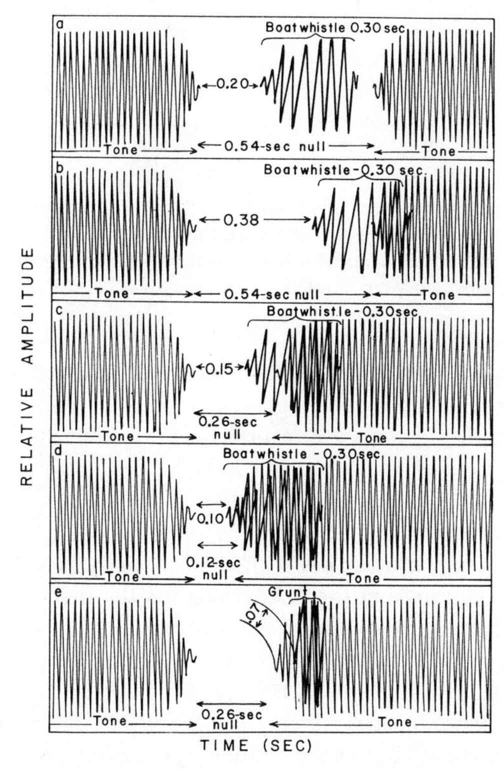

Fig. 19. Diagram of oscillograms showing **(a)** boatwhistle response fitting entirely in a 0.54-sec silent interval, **(b)** boatwhistle beginning later in a 0.54-sec interval and overlapping the start of the subsequent tone, **(c)** boatwhistle beginning in a 0.26-sec silent interval, **(d)** boatwhistle beginning in a 0.12-sec interval, **(e)** grunt response beginning 0.07 sec after start of tone (playback 1T.26).

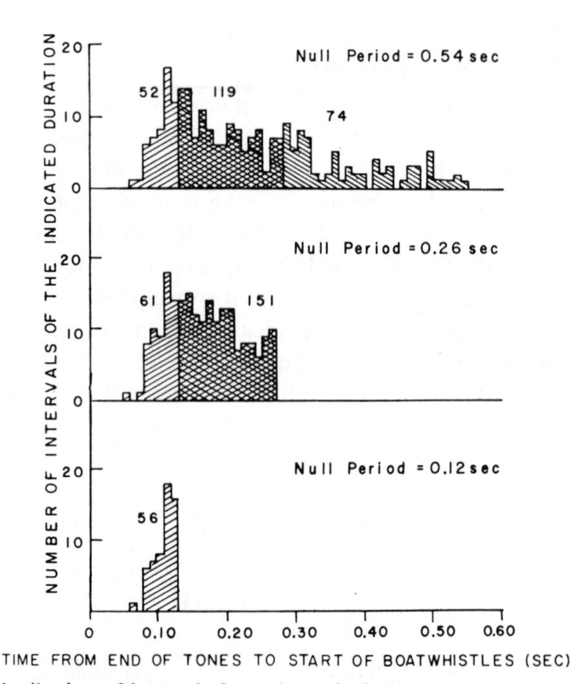

Fig. 20. The distribution of intervals from the end of the 6-sec tones to the onset of the boatwhistle responses which began in the silent intervals, for the three one-tone playbacks (1T.54, 1T.26, 1T.12) of the response-time experiment. The last interval in each histogram contains those calls which were emitted simultaneously with the onset of the succeeding tone.

There was no significant difference in the number of boatwhistles produced during the control and playback periods of 1T.54 (Wilcoxon $T = 22.5$), but a significant decrease in boatwhistle calling (0.05 level) occurred during the playback of 1T.26 and 1T.12. Wilcoxon T values were 8.0 and 7.0, respectively.

The histograms in Fig. 20 show the distributions of intervals from the end of the 6-sec tones to the onset of the boatwhistles which started in the silent intervals. Comparing the histograms for the 0.12- and 0.26-sec silent interval playbacks, it is apparent that only slightly more responses began during the first 0.12 sec of the 0.26-sec silent intervals than during the actual 0.12-sec gaps of the 1T.12 playbacks (61 compared to 56), even though the total number of boatwhistles starting in the 0.26-sec gaps was 212 compared to 56 starting in the 0.12-sec gaps. Rather, most of the additional responses started 0.12–0.26 sec after the tones ended. Similarly, the number of responses starting in the first 0.12 sec of the 0.54-sec silent intervals was approximately the same (52) even though a total of 245 calls began in the 0.54-sec intervals. It can be seen from the histograms that the minimum response time was about 0.08 sec, with response times of 0.12–0.14 sec, corresponding to the histogram modes, being most common.

Generally, playback was on for at least 25 sec (four cycles) before the first boatwhistle was produced, particularly when the null period was only 0.12 sec. Also, quite often the first few sounds produced were grunts, then overlapping boatwhistles (beginning while a tone was playing), then non-overlapping boatwhistles (beginning in the silent intervals). This observation, plus the extremely fast response time, led me to believe that the test fish might have been learning the rhythm of the regular pattern of 6-sec tones followed by small silent intervals. To test this, a series of three patterns with 5-, 6-, and 7-sec tones (Fig. 18) was presented to the test fish the following year. The playback tapes were constructed in the field, without the aid of a high-speed timing device, and the silent intervals between tones turned out to be slightly longer than those of the 1T, regular series.

Of the 274 boatwhistles produced during the ten replicate playbacks of 3T.62, 82% began in the 0.62-sec silent intervals (compared to 86% starting in the 0.54-sec intervals of the analogous regular pattern 1T.54). Similarly, 91% of the calls produced during the 3T.31 playbacks started in the intertone gaps (compared to 93% for the analogous 1T.26 playbacks), and 74% produced during the 3T.18 playbacks began in the 0.18-sec intervals (compared to 43% for 1T.12). In the latter comparison, the difference between 74 and 43% was probably due to the difference in duration of the silent intervals (0.18 *vs.* 0.12 sec). The 0.12-sec interval was approaching the minimum response time for the fish. There was a significant reduction in calling from the control to the playback periods of 3T.31 (Wilcoxon $T = 4.0$) and 3T.18 ($T = 3.0$), but not of 3T.62 ($T = 15.5$).

The histograms in Fig. 21 show the distributions of intervals from the end of the tones to the beginning of the boatwhistles which started in the silent intervals. Comparing the histograms for the 3T.31 and 3T.18 playbacks shows that while there were more responses beginning in the 0.31-sec intervals than in the 0.18-sec gaps, they started later than 0.18 sec. Also, comparing the first 0.31 sec of the histogram for 3T.62 to the histogram for 3T.31 shows that only two of the additional 44 responses in the former began during the first 0.31 sec ($124 + 58 = 182$ compared to $110 + 70 = 180$). The onset of the remaining 42 responses in the 0.62-sec intervals was from 0.31 to 0.62 sec. As with the one-tone series, then, the number of responses beginning in the gaps was greater when the silent interval duration was longer, but the number starting during equivalent time sections of the silent intervals was very similar.

The specific location of all grunt-type sounds was determined in addition to boatwhistles. Of 684 grunts produced during all replicates of the six playback types, only 149 or 22% began in the nulls compared to 81% of the total

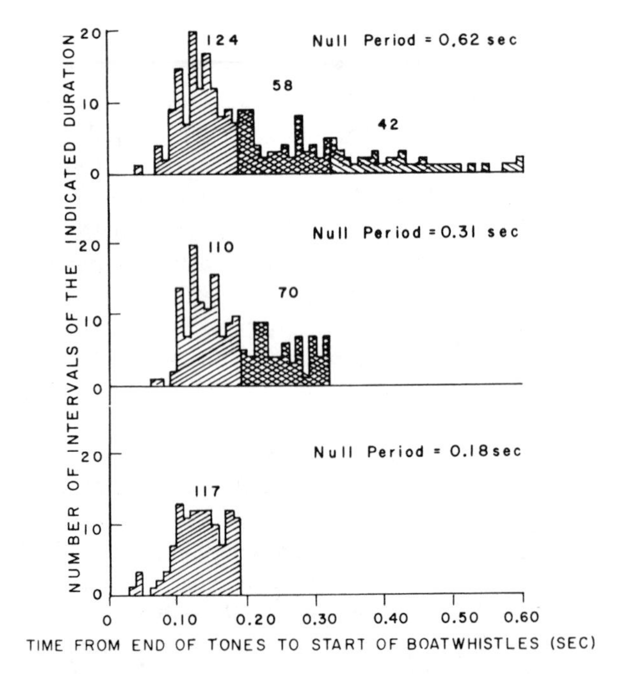

Fig. 21. The distribution of intervals from the end of the tones to the onset of the boatwhistle responses which began in the silent intervals, for the three three-tone playbacks (3T.62, 3T.31, 3T.18) of the response-time experiment. The last interval in each histogram contains those calls which were emitted simultaneoulsy with the onset of the succeeding tone.

1271 boatwhistles. The 237 boatwhistles that did overlap the tones occurred at all points in time from the beginning to the end of the tones. But all except 14 of the 535 overlapping grunt responses began during the first 0.10 sec of the tones and most of these during the first 0.07 sec (Fig. 19e).

IV. DISCUSSION

There is behavioral evidence that calling rate may be a very important parameter for transmitting information vocally in the toadfish. Gray and Winn (1961) found that a male toadfish increased his calling rate when a live, ripe female on a string was pulled past his nest. I witnessed a similar increase in rate during one of the control periods of this investigation. A male had been calling at a rate of about ten boatwhistles per minute; when a female approached to within 2 m of his shelter, he accelerated his calling rate to 25 boatwhistles per minute. The female swam quickly and directly toward the fast-calling male and entered the nest. Several grunts were produced, but no boatwhistles were emitted until the female left the shelter about a minute later.

In this study, it was possible to increase the calling rate of the test fish with many types of playbacks, providing the repetition rate of the playback exceeded a certain value. All parameters of the fish's vocalizations, other than rate, remained the same during the playback as during the preplayback control periods; that is, there was no change in the amplitude, temporal fine-structure, frequency, or duration of the calls. Apparently, a maximum sustained calling rate exists. I will define it as the fastest rate at which the fish can be driven to rhythmically call for long periods of time. Considering the results of all the playback experiments, I estimate this rate to be one call every 3.7–4.3 sec. The highest calling rates should and do occur during those playbacks which have cycle times equal to 3.7–4.3 sec or some submultiple of 3.7–4.3 sec. In the latter case, the fish can skip every other or every third playback sound and still call every 3.7–4.3 sec without overlapping a playback sound.

In the calling-rate experiment, both playbacks were too fast for the fish to call after each cycle for long periods of time. There were more boatwhistles produced during the faster playback, BW24 (one playback sound every 2.5 sec), than during BW20 (one playback sound every 3.0 sec) because the first multiple of the cycle time of BW24 (2.5 sec $\times$ 2 = 5.0 sec) was closer to the preferred cycle time of 3.7–4.3 sec than the first multiple of the cycle time of the slower-rate playback (3.0 sec $\times$ 2 = 6.0 sec). A sustained calling rhythm occurred for longer periods of time with the faster playback.

Winn (1967) showed, for constant-interval boatwhistle playbacks, that once the playback sound was broadcast the fish would not overlap the trailing edge of the sound with a response. Instead, the response occurred after the playback boatwhistle ended. Similarly, in the boatwhistle-pattern experiments discussed here, where the playbacks consisted of bursts of boatwhistles, many more responses fell in the longer silent intervals separating the bursts than in the intraburst intervals or simultaneously with the playback sounds (overlaps). This resulted in a change in the gross temporal calling pattern of the test fish from the control to the playback periods. There was no evidence, however, for mimicry of the playback patterns. The fact that response locations could be accurately predicted for most playback types (as shown by the narrow peaks in the histograms) was due only to the fish spacing their responses between playback bursts to avoid overlapping the playback sounds. The rhythmic response pattern was a function, then, of the rhythmic playback pattern or, more precisely, the pattern of silent intervals between playback bursts. There was no entrainment or carryover effect of the pattern after the playback ended. Also contributing to the change in histogram shape from control to playback was the increase in calling during the playback periods of all the 200 boatwhistle per 3 min playbacks and all but two of the 90 boatwhistle per 3 min playbacks. None of the six slow-rate playbacks (30 boatwhistles per 3 min) caused an increase in calling (or a change in calling pattern) regardless of the playback pattern type.

In the boatwhistle-pattern experiment, the test fish reacted to the six fast-rate playbacks (200 boatwhistles per 3 min) by calling infrequently for the first 30 sec. Most responses that did occur during this time overlapped the playback bursts. Fewer overlaps occurred as playback continued, suggesting that the fish may have detected the rhythm of the playback patterns. Also, three of the fast playbacks increased grunting, and half of the grunts came during the first minute of the playback periods before boatwhistle calling had reached a peak. It is possible that the grunts represented potential boatwhistles that were cut short by playback bursts before a calling rhythm was established.

Each cycle of the six, complex boatwhistle patterns (series 5-BW and 6-BW) contained two or more different-length bursts. The intervals between bursts were constant for each playback type. With all of these playbacks except 5-BW200, the boatwhistle responses were distributed equally after the playback bursts regardless of burst length. In 5-BW200, however, over 70% of the responses came after the 2.81-sec four-boatwhistle bursts and less than 30% after the 0.38-sec single boatwhistle even though the interburst intervals were the same (0.67 sec). The playback caused an increase in boatwhistle calling, but because of the playback pattern the fish were suppressed for a longer time by the four-boatwhistle bursts than by the single boatwhistle. The longer bursts provided nearly a 3-sec rest; thus, even if the test fish had called after the preceding single playback sound, he could easily call again after the four-boatwhistle burst. But if he had called after the four-burst, it would have been necessary to respond again in less than 1 sec to call after the next single playback sound. Since this was almost impossible, once a sound was placed after a four-burst, it was likely that the next response would occur there also, and the calling pattern was consequently established. Thus, what appeared initially to be patterning can be explained simply by a limit as to how fast the fish can call two times in a row.

It was demonstrated by the cycle-time experiment that toadfish, stimulated by sound playback, have an optimum calling rate of 14–16 sounds per minute (one sound every 3.7–4.3 sec). When the playback cycle time was 4.1 sec, a response occurred after nearly every cycle of the five-boatwhistle-burst playback (Fig. 13). Thus, calling rate was finely tuned to the rhythm of playback and alternation was maximum. However, when playback cycle time was less then 3.5 sec, the number of responses occurring after each playback cycle decreased while the number after every other cycle increased. When the cycle time was only 2.4 sec, many more responses fell after every other cycle (4.8

sec) than after every cycle since the former would allow a calling rate closer to the optimum sustained rate of 14–16 sounds per minute.

Winn (this volume) found that 200-Hz-tone bursts of approximately the same duration as natural boatwhistles caused an increase in calling (as boatwhistle playbacks did) when played at a constant stimulatory rate. My tone-pattern experiment, discussed in this chapter, was designed to test the effect of longer and variable duration tones in both simple and complex patterns. The fish responded by increasing their calling rates just as much as when boatwhistle playbacks with similar cycle times were played to them. The durations of the 200-Hz tones (1.0–3.0 sec) were not even similar to the duration of an average boatwhistle (less than 0.4 sec). The 2.60-sec tone playback (TP3) with a 4.10-sec cycle time had the same stimulatory effect as the 2.60-sec burst of five boatwhistles (playback 5-BW4.1) with a 4.10-sec cycle time. The distribution of responses in the intertone intervals of the tone-pattern playback TP1, where each cycle contained a short (1.0 sec) and a long (3.0 sec) tone, was the same as the distribution of responses in the interburst intervals of the boatwhistle playback 5-BW200, composed of a single boatwhistle and a four-boatwhistle burst. About two thirds of all responses fell after the long tones or bursts.

The fact that toadfish were stimulated by playbacks of both real sounds and 200-Hz tones of various durations suggests that it is not necessary for the acoustic stimulus to be very specific. But, while the playback signal units do not have to be highly specific, the repetition rate at which they are presented must be above a certain threshold value. Winn (1967) found that a rate of ten sounds per minute was not stimulatory but a rate of 18 sounds per minute or greater was. In the cycle-time experiment, I determined a maximum sustained calling rate of 14–16 sounds per minute. This explains why Winn did not observe a difference in the stimulatory effect of playbacks with rates of 18, 26, and 36 sounds per minute; all were stimulatory and above the maximum capable sustained calling rate of the toadfish.

Fish (1954) first associated the boatwhistle call with the inception of breeding season. She also observed that more than one female contributed to the accumulation of eggs since partially spent males, still producing boatwhistles, were found guarding nests of eggs in several stages of development. When silent males were found on nests, the eggs were in later stages of development and the males were completely spent. Gray and Winn (1961) hypothesized that the biological significance or function of the boatwhistle "is an attractive stimulus to females that are ready to lay eggs." This was based on experiments where males responded with boatwhistles when ripe females were passed in front of their nests; males previously calling increased their calling rates. Winn (this volume) reports other evidence that the sound attracts females. But this still does not explain why males are stimulated to call faster when presented with sound playbacks of boatwhistles that are only produced by other males.

One possible hypothesis can be developed by combining some previously discussed observations. Males increased their calling rates only when the playbacks exceeded a threshold rate (about 12–14 sounds per minute); males started boatwhistling or accelerated their calling on the approach of a ripe female. If numerous males are located within hearing range of each other (as they were in the sound playback experiments described here) and are competing for females, and if increased or accelerated calling functions as an attractive stimulus to a passing female, then a single male that has observed the near presence of a female (and hence increased its calling rate) may indicate the female's presence to nearby males and in turn cause them to call faster. Direct, precisely timed calling back and forth by two male fish does not occur, however, based on the results of the antiphony experiment.

Winn (1967) showed that boatwhistle calling suggestive of antiphony with playback could be explained by chance alternation. The playback tapes were constructed with equal intervals between sounds; toadfish calling is often regular, thus many alternations could occur by chance alone. He suggested that in the natural environment one fish calling at a regular rate could stimulate another fish to increase its calling (also at a regular rate) and that alternations would occur by chance. The system is enhanced by the tendency of the fish to avoid overlapping each other's sounds. The antiphony experiment reported on here was designed to test Winn's hypothesis.

To determine if regular alternation would still occur when a test fish was given the opportunity to control the playback rhythm by his response speed, each call the animal produced was followed by a delayed playback of that call instead of a previously constructed equal-interval playback. When the playback delay time was 1.55 sec, calling rate increased over the preplayback control rate significantly more than it did when the delay time was 3.10 sec. This was probably because the maximum sustained threshold calling rate of 14–16 sounds per minute (one call every 3.7–4.3 sec) could not be attained when the playback delay was 3.10 sec without the fish responding to each playback sound within 0.60–1.2 sec. While such response times were certainly possible (demonstrated by the response-time experiment), most of the responses still fell about 1.5–2.0 sec after the playback sounds, as they similarly did when the playback delays were 1.55 and 2.05 sec. This rhythm of responding once to each playback signal within 1.5–2.0 sec was not maintained for more than a few cycles, though, when the playback delay was 3.10 sec. Instead, one third of the playback signals were followed by two boatwhistles, the second being emitted before the delayed playback of the first was projected. Such double responses enabled a fish to more closely approach the maximum stimulatory threshold calling rate of 14–16 sounds per minute.

When the delay time was 2.05 sec, an antiphonal response rhythm to the playback occasionally seemed to occur. The rhythm, however, can be explained by alternation other than true antiphony. The fish were stimulated by the playback to call at the maximum sustained rate of 14–16 sounds per minute (one sound every 3.7–4.3 sec). This rate happened to coincide with the rate established by the sum of the playback delay time (2.05 sec) plus the average response time (1.5–2.0 sec), which was the same for all playback delays; the sum equals 3.55–4.05 sec. One call every 3.55–4.05 sec equals a calling rate of 15–17 sounds per minute, which is about the same as the maximum sustained calling rate.

A similar phenomenon probably occurs in the natural situation. Two nearby calling fish could stimulate each other to call at or near the maximum sustained rate (one call about every 4 sec); a phasing could at least temporarily result where the sounds of two animals would occur 2 sec apart; the phasing or rhythm could be maintained by suppression (inhibition) during part of the cycle (to avoid overlapping each other's sounds). The resulting alternation could easily be mistaken for antiphony. It was shown by the cycle-time experiment that the phasing could be controlled to within 0.10 sec when the fish were presented a fixed-rate playback with a rate equivalent to the sustained threshold calling rate. The precision of responding was good, and rhythmic alternation resulted. When the fish were not driven by a fixed-rate, constant-interval playback, but instead could select their own calling rates by responding within a certain time after the playback signals, precision was not good. Consequently, alternation was not rhythmic.

A convenient means of estimating the auditory reaction time of the toadfish resulted from my learning that the fish would not call while the tones were playing during the tone-pattern experiment. Rather, they called during the silent intertone intervals. Response time, then, could be measured from the end of the tones to the beginning of the boatwhistles. However, when the playback tones were less than 3 sec and the silent intervals between tones longer than 1 sec, the fish were stimulated to call faster (at the threshold rate of 14–16 sounds per minute) and to emit their calls near the center of the silent intervals, thus being attuned to the playback rhythm. This would not provide an estimate of the minimum response time. When the playback tone duration was increased to over 5 sec and the silent intertone intervals were reduced to less than 1 sec, the fish were suppressed for a longer period of time and could not maintain the threshold calling rate even by responding in each silent interval. Under these conditions, the fish would respond very quickly in the silent intervals when the tones ended.

Minimum auditory times determined by this method were 0.08 sec, with response times of 0.12–0.14 sec being most common. There was no indication that the fish were able to estimate the duration of the silent intervals and adjust their response times accordingly. Three different playbacks were utilized consisting of a 6-sec tone followed by intertone intervals of either 0.54 sec, 0.26 sec, or 0.12 sec; the distribution of responses was the same within equivalent time sections of the silent intervals. If an adjustment to the silent-interval duration had occurred, there should have been either more responses in the short 0.12-sec silent intervals than in the equivalent 0.12-sec section of the 0.26- and 0.54-sec intervals, or the modes of the histograms for response times should have had different locations for the three different silent intervals. Neither was the case. It seems that at any given time following an inhibitory tone there is a certain probability that a response will occur. If this time is eliminated by the start of a subsequent inhibitory tone, that response will simply not be emitted. The limit, of course, is total inhibition by a continuous tone.

There was no difference in the response time with the regular and irregular playback patterns or in the number and distribution of boatwhistle and grunt responses. About 90% of all the calls produced began in the silent intervals even when the intervals were only 0.26–0.32 sec long. In contrast to boatwhistles, grunts overlapped the tones and increased in number as the silent-interval duration decreased. This, plus the fact that nearly all the overlaps occurred within 0.07 sec from the onset of the tones, suggests that the grunts may have actually been the beginning of boatwhistles which were initiated by the nervous system during the final moments of the silent intervals. Upon hearing the start of the next tone, the fish cut the potential boatwhistles short, resulting in so-called grunts. Oscillograms show that the first eight to ten cycles of a boatwhistle are identical to a grunt.

It is interesting to compare this work on the vocal communication system of the toadfish, *Opsanus tau*, to studies done on other species of fish and other animal groups. Moulton (1956) incited sea robins (*Prionotus* spp.) to produce the staccato call by playing back recordings of sea robins' staccato calls. In addition to the natural calls, crude and distorted artificial staccato sounds, generated by an audio oscillator, caused a similar increase in calling, indicating once again that the acoustic stimulus may not have to be highly specific.

The functional significance of a fish sound was experimentally demonstrated by Tavolga (1958*b*) by playing back various sounds to male and female gobiid fish, *Bathygobius soporator*. He found that the sounds were related to courtship. Males increased their activity and probably their level of responsiveness when presented with male sounds, thus increasing the probability of several animals competing for a female. Playback consisted of a continuous series of natural grunts with a constant repetition rate (about one sound per second) and artificially generated sounds. Positive results were obtained with both playbacks even though the artificial sounds differed in duration, amplitude, and frequency from the natural sounds. Although the calling rate of the male fish was extremely variable, it increased when a gravid female approached. Recall that male toadfish also increased their calling rates when females approached the nests.

A preference experiment was performed by Delco (1960) to determine the possibility of fish sounds being used as a species recognition device. He tested the discriminating ability of two cyprinid fishes, *Notropis venustus* and *N. lutrensis*, to their own sounds. Ripe females and males of both species were presented recorded calls of their own and other species simultaneously at opposite ends of a tank. Males differentiated recorded calls by females of their own species and approached the correct speaker. The sounds of the two species differed in call duration, frequency, and trill rate. No experiments were performed, however, to determine what aspects of the sounds enabled the discrimination.

Several other experimenters have observed calling-rate changes under various conditions. Winn and Stout (1960) showed that a group of male satinfin shiners, *Notropis analostanus*, injected with testosterone produced many more sounds than either a group injected with sesame oil or a natural control group, indicating that the sound may be related to reproductive activities. Stout (1966) showed that territorial males produced single knocks when they chased intruding males. During more vigorous aggressive displays between males, the single knocks were united into a rapid series of knocks; temporal pattern or repetition rate changed. Stout determined the functions of the various sounds by playing sounds back and observing the movements and reactions of the fish to each other.

In contrast to toadfish and most other fish which make only one or two types of sounds, some anurans, birds, and insects produce many kinds. Frogs make at least five different types; their functions have been well described by Bogert (1960) and Capranica (1966, 1968). The former author classed them as mating, territorial, release, warning, and distress sounds. Capranica found the vocal repertoire of the bullfrog, *Rana catesbeiana*, to consist of seven distinct types of sounds with little variation in call structure among different animals from the same or even different geographic locations. The seven stereotyped sounds were distinguishable by gross temporal pattern, fine temporal structure, and spectral distribution of energy. Three different calls were associated with territorial defense, one made only by males, the second only by females, and the third by both sexes. The mating call, made only by males (as with toadfish), was produced in response to natural or synthetic mating calls. Capranica suggested that the call may function during the breeding season to congregate other males for chorusing and possibly to attract females or initiate their seasonal reproductive cycle. In addition, the call may serve to maintain the geographic extent of the population by species identification, as it is also produced outside of the mating season. The male frogs reliably responded with the mating call to natural and synthetic mating calls presented at all times of the year. Toadfish, to the contrary, neither produced boatwhistles nor responded to sound playbacks except during the breeding season. Also in contrast to toadfish, the frogs would only respond to sounds very similar to their own. In fact, when presented sounds of 33 different species they responded only to the calls of other bullfrogs. This supports Blair's (1958) hypothesis that the calls of male frogs act as a mechanism promoting reproductive isolation.

To determine what parameters of the frog sounds were responsible for this species-specific response, Capranica played back synthetic calls, varying in frequency, amplitude, and temporal fine-structure. A certain complex frequency–amplitude relationship was necessary to elicit a response. However, all the playbacks had fixed rates of one 0.8-sec croak beginning every 1.5 sec; it would be interesting to vary this parameter also. Capranica (1968) concluded that the temporal features, in addition to spectral characteristics, are probably of considerable importance to the bullfrog in recognizing the different sounds.

While the study of fish sounds began only in the late 1940s, research on vocal communication in birds has been conducted for many decades. Hundreds of papers have written on the subject, including recent reviews by Armstrong (1963), Busnel (1963), Lanyon and Tavolga (1960), and Thorpe (1961). Sounds produced by birds are extremely complex and diverse, and undoubtedly information is carried by more than one parameter of the sounds. Some species produce many different songs with variations in frequency, amplitude, and temporal pattern. For example, Borror (1961) showed that the finch makes at least 13 different theme songs with 187 variations. Intraspecific variation apparently occurs in nearly all forms. No studies to date have been concerned with the ontogeny of fish sounds, but there are numerous reports on the development of bird song. Part of the basic song develops in young birds raised without ever hearing adult birds of their own species. Rice and Thompson (1968) recently demonstrated this with the indigo bunting. The general shape of the units ("figures") making up the sounds and their spacing, repetition, and frequency range were similar for birds raised in isolation to those of wild birds. Learning seemed to merely refine the basic pattern. Lemon and Scott (1960) showed similar results with the cardinal. Simple units of the song patterns developed in isolated birds, but the complex parts had to be learned.

Even though considerable variation exists in the songs produced by different individuals of a species, the song of any particular individual is reasonably fixed in its physical structure and organization. When both sexes make sounds, such specificity aids in individual recognition, contributing parental, filial, and social information (Busnel, 1968). In some species of shrikes, both males and females have a large repertoire. They develop a series of duet patterns which they presumably use for mate recognition when out of each other's sight. These duet patterns are distinguished from those of other pairs in the neighborhood by a very precise and exactly maintained time interval (varying by less than 1.5 msec) between the calls of the two sexes (Thorpe, 1963). Thus it appears that the gross temporal parameter is important for transmitting information vocally in certain birds as well as fishes. Other parameters of the sounds (frequency, harmonic structure, amplitude, temporal fine-structure) may play a less significant role in intraspecific information transmission than might be expected from the complexity of the sounds. As long as the signal has a certain basic composition, conspecific information may be transmitted by calling rate, call duration, and response time. Variation in the acoustic parameters of the sound may be more important in interspecific recognition (for example, a territorial function) or lack of interspecific recognition (preventing mating in closely related species).

In contrast to fish sounds, where function has not been adequately demonstrated except for the few cases previously mentioned, the functional significance of many bird sounds has been well documented. Collias (1960) classified the sounds as related to group movements, food finding, avoidance of enemies, and reproduction (subdivided into sexual and parental phases). The "bobwhite" call is a sexual call produced by an unmated quail during the breeding season (Stoddard, 1931). Once the male finds a mate, calling stops almost entirely (similar to toadfish); if the female is removed, the male will resume its calling after several hours. Stokes (1967) showed that, ordinarily, the male quail gave his "bobwhite" call in any direction at four to five calls per minute. On hearing the "hoy-ee" call of a female, the male faced the direction of the calling hen and increased his calling rate to eight to nine calls per minute one call coming right after another without a break. This increased calling by the male on sensing a female is remarkably similar to the increased boatwhistle calling of a male toadfish on the approach of a female. Also, playback of the "hoy-ee" call caused an increase in "bobwhite" calling of the male quail.

The ability to memorize and imitate the signal characteristics of other species has only been demonstrated for birds (Armstrong, 1963) and porpoises (Lilly, 1962). Mockingbirds and starlings regularly include notes and phrases of many other species in their songs. Indian mynah birds and parrots can imitate human speech and various noises without too much practice. Thorpe and North (1965) suggested that this extreme imitative ability occurs where the main function of the song is to provide for social recognition and cohesion rather than for territorial defense, as it is most pronounced in tropic and subtropic environments where dense foliage often limits visual contact.

Male and female shrikes, *Laniarius aethipicus major*, each had their own exclusive songs which were not normally used by their mates. However, when the partners were absent the remaining member uttered the call of the missing partner, resulting in the latter's return. Hence, one major function of the imitative ability is to establish and strengthen the individual pair bond.

Duet calling between the male and female of a pair has been demonstrated for many species of birds. Thorpe and North (1965) found that it occurs in tropical barbets, fantail warblers, and shrikes. One pair of shrikes produced 12 different duet patterns; another pair made 17 in a single day. In all these groups, the songs of the male and female are different—the sexes alternate antiphonally with a very precise and exactly maintained time interval between each contribution. Stokes (1967) showed that the male "bobwhite" response to the playback of female "hoy-ee" sounds also tended to be antiphonal, with the first syllable of the "bobwhite" call mingling with the end of the "hoy-ee" call. In nature, the separation call of the female, variously described as "hoy," "hoy-poo," and "koi-lee," often elicited a similar call (antiphonally produced) from a displaced male (Stokes and Williams, 1968). A more common form of antiphonal calling in the bobwhite was that of an unmated male and a hen that had become separated from her mate. The female's separation call ("hoy") would be followed antiphonally by a "bobwhite" call from the unmated male, overlapping the "hoy" sound. There was very little variation in the mean response time between the onset of the two sounds. If the male reacted to the very onset of the female's call, an occasional antiphonal response to an inappropriate sound or call should occur. Stokes and Williams suggested that this was prevented by the male not responding to the first few female separation calls and then responding only softly until the female's rhythm was established. A similar phenomenon occurred in many of the toadfish playback experiments.

Diamond and Terborgh (1968) divided synchronized duetting into three categories, restricting the term "antiphonal" to the type of duet in which the male and female sing alternately, with the call of the second (generally the female) beginning precisely upon cessation of the call of the first (generally the male). In the second category, the pair sing different phrases simultaneously; the synchronization of the phrases may be very exact, with the female beginning her sound only a fraction of a second after the male. (In the bobwhite, this order was reversed.) The third category is typified by the male and female calling in unison. A requirement of all categories is the time synchronization of the sounds.

Synchronous calling has not been demonstrated between male birds, except for one special case of the unison type (two male blue-backed manakin birds perch next to each other, facing a female, and call toward her in unison). So-called countersinging (not synchronous) does occur between males, however. When pauses between successive phrases of a territorial song are longer than the song phrases, males on adjacent territories may tend to sing alternately (Diamond and Terborgh, 1968). Male–male countersinging differs from male–female synchronous calling in that the songs of the two males are identical and not well coordinated. (This was also found with male toadfish. A fast-calling fish stimulated its neighbors to call faster, but there was no precise synchronization among the animals.)

As with birds, many good papers and reviews have been written on sound production and reception in insects (Alexander, 1960, 1967; Busnel, 1963; Pierce, 1948). Some insects use sound in their behavioral displays in ways very similar to birds. In most cases, only male insects make the sounds, and the signals are purely intraspecific. Spooner (1968) experimentally demonstrated movements of katydids upon hearing sounds of conspecific species. Random movement and frequent turning begins with the sound and continues until the sound ends. Spooner suggested that this could result, in natural populations, in the spacing of individuals. In singing insects, an immediate function of at least certain sounds is to bring males and females together during the breeding season. This is especially true where population densities are low; effective breeding can occur only when a special mechanism (such as sound production) aids in the pair formation (Spooner, 1968; Shaw, 1968). The males' calling (which may be produced for hours at a time) attracts the sexually responsive females. In some species, it also attracts males and stimulates calling in other males (Alexander, 1960).

Long-term, rhythmic alternation between neighboring male katydids was shown by Shaw (1968); eventually, sexually responsive females were attracted to the area. Alexander (1967) suggested that alternation may assist in the formation of groups of males (by congregation and/or spacing). Shaw compared the periods between chirps of soloing and alternating male katydids and found that they were greater during alternate singing. The acoustic interaction between the two individuals resulted in an entrainment of each katydid at a rate slower than the solo calling rate because of inhibition by the acoustic stimulus (chirp of the other katydid). Playback experiments confirmed this observation—calling could be slowed by presentation of electronically produced imitation chirps at fixed and continuously variable rates. Increasing the intervals between the stimulus chirp and the preceding responder's chirp increased the effectiveness of the acoustic stimulus in slowing the chirp rate of the animal. There was, however, a limit to the amount of delay that could be imposed; at this point, the katydid would respond before the end of the stimulus chirp.

Calling rates of toadfish could not be slowed by playback as was done with katydids. To the contrary, calling was always greater during fast-rate playbacks than during the preplayback control periods. This was true even when the fish had the opportunity to actually control their own rate of calling and still maintain a precise rhythm with the playback signals, as in the antiphony experiment. The unique manner of playback presentation in the antiphony experiment permitted a true test of the precision of response rhythm that might be expected between two neighboring animals. Other experimenters (working with insects, birds, and fish) presented either fixed-rate playbacks (thus forcing a response rhythm and increasing the chances of alternation), or continuously variable playbacks (decreasing the chances of long-term alternation). However, making the interval constant between the animal's sound and the subsequent playback signal allows the test animal to select its own calling rate (by its response rate) and still maintain a precise response rhythm. By varying this playback delay interval in different experiments, one can obtain information on the limits of alternation and the precision of response timing. The results of the antiphony experiments on toadfish, using this type of playback, showed that the precision of the response rhythm was much less than might have been concluded on the basis of a fixed-rate playback. This was particularly so when the fixed-rate playback happened to have a cycle time equivalent to the period between responses produced by a fish which was stimulated to call at the maximum threshold rate. Compare, for example, the width of the 5-BW4.1 histogram in Fig. 13 (fixed-rate playback) to the BW2.05 playback period histogram in Fig. 14 (response-controlled playback). The fish could call at the same rate (the threshold rate) during both playbacks, yet the precision of response was much greater with the fixed-rate (driving) playback (resulting in the very narrow-peaked histogram) than with the response-controlled playback.

In Shaw's experiments, the timing precision of test katydids in responding to a playback, or to a second (follower) katydid of a male pair, varied by less than ± 0.05 sec between responses even when the interval between playback signals was variable. Such precise timing was never encountered with toadfish; responses fell at all locations within the intervals between playback signals, especially when the playback pattern was complex. Also, precision alternation between calling males has not been shown for birds and not experimentally demonstrated for anurans.

Spooner (1968) pointed out (for insects) that not only are male songs species specific (particularly in species where both males and females make sound), but the timing of the female response is characteristically specific among species, particularly where heterospecific male songs are similar enough to cause confusion. This same feature of precision timing within a species (or male–female pair) was present in antiphonal calling birds.

It may turn out that true antiphonal calling will be limited to those special cases where male–female pair formation occurs, with male–male alternation resulting merely from mutual inhibition by each other's sounds.

In the latter case, alternation is more of a forced phenomenon (related to rivalry), while alternation between sexes is voluntary.

Jones (1966*a*) observed this effect in bush crickets, *Pholidoptera girseoaptera.* Mutual excitation appeared to be related to rivalry behavior. In addition to the males' increasing their calling rates, mutual inhibition was an important factor in determining a pattern of alternation. The precision of alternation was much greater, however, than in toadfish; the chirps of one cricket were either centrally placed in the interval between the chirps of the second cricket or were in near perfect synchrony (unison). The change from perfect alternation to synchrony was abrupt with no near overlaps. Even when the calling rates of the two males differed, there was still no transition from alternation to synchrony. Instead, the slower-calling insect would alternate for a few cycles, then stop calling for several cycles, and finally start calling again—either in synchrony or alternation.

Jones (1966*b*) played back various types of sounds to crickets and obtained some results which are quite similar to the results of the toadfish experiments. The male crickets could be induced to sing in alternation with many types of sounds, including a Galton whistle, hissing sounds, and pure tones from a signal generator. The physical structure of the stimulus did not have to be very specific. Actively chirping insects would alternate with certain high-rate playbacks, but less active individuals would skip one or two playback sounds and respond during the following silent intervals (as toadfish did). Playbacks of 5-, 1-, and 0.1-sec duration to the crickets all caused increased calling. Cricket chirp rate was suppressed (as with toadfish) when continuous playback signals were used, except in certain experiments with pairs of males. For normally excitatory signals, an increase in playback intensity from 40 to 70 db (re 0.0002 μbar) resulted in an even greater stimulatory effect. It was similarly demonstrated in a threshold experiment on toadfish (Fish, 1969) that the fish were stimulated to call at faster rates by higher-level playbacks. Thus, intensity, too, may play a role in information transmission, possibly relating to competition. A louder signal could indicate the presence of a close rival and hence stimulate faster calling.

While there is no reason to believe that one acoustic parameter of the signal can be singled out as the only one necessary or useful for transmitting information in a particular behavioral situation, calling rate and other temporal factors do seem to be of paramount importance. This was shown, with examples for fish, birds, and insects. Frequency may have its most important function in species recognition and initial recognition of a particular behavioral situation (especially in birds that make several calls which vary in frequency), with information about levels of excitation within the situation being transmitted by rate.

The least variable parameter of fish, bird, or insect sounds is generally the temporal fine-structure, that is, the pulsed repetition of the small units which make up the calls, chirps, etc. These individual pulses are not often discernible by the human ear. The pulse rate is a close expression of the mechanism responsible for the sound; in insects, it is temperature dependent (Frings and Frings, 1962).

I can only roughly compare the minimum and mean auditory reaction times of the toadfish (0.08 sec and 0.12–0.14 sec) with other animals since data are lacking for most. No other similar experiments have been performed on fish. Estimates for birds vary with different species. Stokes and Williams (1968) demonstrated reaction times mostly between 0.30 and 0.50 sec in antiphonal calling quail. Mean reaction times of 0.09 sec and minimum times of 0.07 sec were found in the antiphonal exchanges of a tropical shrike (Thorpe, 1963). Jones (1966*a*) found that mutual inhibition between singing male bush crickets (with a 0.05-sec inhibitory reaction time) played a major role in determining the timing of the chirps by each male. I presented the same playbacks used in the toadfish experiments to humans and got minimum response times of 0.14–0.15 sec, with most responses starting about 0.20 sec after the end of the tones.

The fast response time of the toadfish, *Opsanus tau*, can probably be attributed to its sophisticated sound-producing mechanism. Its sounds are produced by contracting special striated muscles which are completely attached (bilaterally) to the swimbladder (Tavolga, 1964). These muscles are able to contract at a very high frequency (greater than 100 per second) without mechanical fusion (Gainer and Klancher, 1965). They have an average contraction time of 5 msec, an average relaxation time of 8 msec, and an interval between the muscle action potential and the onset of contraction of 0.5 msec. The swimbladder has, in addition, multiple innervation along its entire length, with nerve endings spaced about 100 μ apart. Simultaneous excitation of all the nerves leading to the sonic muscles would evoke simultaneous and distributed action potentials throughout, so that twitch frequency would not be limited by low-velocity conduction of individual muscle fibers. Sounds could be produced extremely rapidly once the swimbladder received the necessary stimulation.

ACKNOWLEDGMENTS

I am grateful to Dr. Howard E. Winn for his counsel and guidance on much of this work and his assistance in the preparation of the manuscript. I also thank Dr. Frank T. Dietz, Dr. H. Perry Jeffries, and especially Professor Robert S. Hass for all their help, and Dr. Michael Salmon for reviewing the manuscript. Mrs. Deborah Kennedy drew the original figures. The

Chesapeake Biological Laboratory. Solomons, Maryland, provided the use of their facilities for the field studieds. Special thanks are due my wife, Nancy, for assisting with all of the field experiments, data analyses, and preparation of the manuscript.

REFERENCES

Alexander, R. D., 1960, Sound communication in Orthoptera and Cicadidae, *in* "Animal Sounds and Communication" (W. E. Lanyon and W. N. Tavolga, eds.) pp. 38–92, Publ. No. 7, Am. Inst. Biol. Sci., Washington, D.C.

Alexander, R. D., 1967, Acoustical communication in arthropods, *Ann. Rev. Entomol.* **12:** 195.

Armstrong, E. A., 1963, "A Study of Bird Song," Oxford University Press, London.

Blair, W. F., 1958*a*, Call difference as an isolation mechanism in Florida species of hylid frogs, *Quart. J. Fla. Acad. Sci.* **21:** 32.

Blair, W. F., 1958*b*, Mating call in the speciation of anuran amphibians, *Am. Naturalist* **92:** 27.

Bogert, C. M., 1960, The influence of sound on the behavior of amphibians and reptiles, *in* "Animal Sounds and Communication" (W. E. Lanyon and W. N. Tavolga, eds.) pp. 137–320, Publ. No. 7, Am. Inst. Biol. Sci., Washington, D. C.

Borror, D. J., 1961, Intraspecific variation in passerine bird songs, *The Wilson Bull.* **73:** 57.

Busnel, R.-G., 1963, "Acoustic Behaviour of Animals," Elsevier Publ. Co., Amsterdam.

Busnel, R.-G., 1968, Acoustic communication, *in* "Animal Communication" (T. A. Sebeok, ed.) pp. 127–153, Indiana University Press, Bloomington, Indiana.

Capranica, R. R., 1966, Vocal response of the bullfrog to natural and synthetic mating calls, *J. Acoust. Soc. Am.* **40:** 1131.

Capranica, R. R., 1968, The vocal repertoire of the bullfrog (*Rana catesbeiana*), *Behaviour* **31:** 302.

Collias, N. E., 1960, An ecological and functional classification of animal sounds, *in* "Animal Sounds and Communication" (W. E. Lanyon and W. N. Tavolga, eds.) pp. 368–391, Publ. No. 7, Am. Inst. Biol. Sci., Washington, D. C.

Delco, E. A., Jr., 1960, Sound discrimination by males of two cyprinid fishes, *Texas J. Sci.* **12:** 48.

Diamond, J. M., and Terborgh, J. W., 1968, Dual singing by New Guinea birds, *The Auk* **85:** 62.

Fish, J. F., 1969, The effect of sound playback on the toadfish (*Opsanus tau*), Ph. D. thesis, University of Rhode Island, *Dissertation Abs.* **31:** No. 2, Abst. No. 70–14, #148.

Fish, M. P., 1954, The character and significance of sound production among fishes of the western North Atlantic, *Bull. Bingham Oceanogr. Coll.* **14:** 1.

Frings, H., and Frings, M., 1962, Effects of temperature on the ordinary song of the common meadow grasshopper, *Orchelimum vulgare* (Orthoptera, Tettigoniidae), *J. Exptl. Zool.* **151:** 33.

Gainer, H., and Klancher, J. E., 1965, Neuromuscular junctions in a fast-contracting fish muscle, *Comp. Biochem. Physiol.* **15:** 159.

Gray, G.-A., and Winn, H. E., 1961, Reproductive ecology and sound production of the toadfish, *Opsanus tau*, *Ecology* **42:** 274.

Gudger, E. W., 1910, Habits and life history of the toadfish (*Opsanus tau*), *Bull. U.S. Bur. Fish.* (1908) **28:** 1071.

Jones, M. D. R., 1966*a*, The acoustic behaviour of the bush cricket, *Pholidoptera griseoaptera*. Alternation, synchronism, and rivalry between males, *J. Exptl. Biol.* **45:** 15.

Jones, M. D. R., 1966*b*, The acoustic behaviour of the bush cricket, *Pholidoptera griseoaptera*. Interaction with artificial sound signals, *J. Exptl. Biol.* **45:** 31.

Lanyon, W. E., and Tavolga, W. N., eds., 1960, "Animal Sounds and Communication," Publ. No. 7, Am. Inst. Biol. Sci., Washington, D.C.

Lemon, R., and Scott, D., 1960, On the development of song in young cardinals, *Can. J. Zool.* **44:** 191.

Lilly, J. C., 1962, Vocal behavior of the bottlenose dolphin, *Proc. Am. Phil. Soc.* **106:** 520.

Moulton, J. M., 1956, Influencing the calling of sea robins (*Prionotus* spp.) with sound, *Biol. Bull.* **114:** 393.

Pierce, G. W., 1948, "The Songs of Insects," Harvard University Press, Cambridge, Mass.

Rice, J. O., and Thompson, W. L., 1968, Song development in the indigo bunting, *Anim. Behav.* **16:** 462.

Shaw, K. C., 1968, An analysis of the phonoresponse of males of the true katydid, *Pterophylla camellifolia* (Fabricius), *Behaviour* **31:** 203.

Spooner, J. D., 1968, Pair-forming acoustic systems of phaneropterine katydids (orthoptera, Tettigoniidae), *Anim. Behav.* **16:** 197.

Stoddard, H. L., 1931, "The Bobwhite Quail," Charles Scribner's Sons, New York.

Stokes, A. W., 1967, Behavior of the bobwhite, *Colinus virginianus*, *The Auk* **84:** 1.

Stokes, A. W., and Williams, H. W., 1968, Antiphonal calling in quail, *The Auk* **85:** 83.

Stout, J. F., 1966, Sound communication in fishes with special reference to *Notropis analostanus*, 1966 Conference on Biological Sonar and Diving Mammals, Stanford Research Institute, Menlo Park, Calif.

Tavolga, W. N., 1958*a*, Underwater sounds produced by two species of toadfish, *Opsanus tau* and *Opsanus beta*, *Bull. Marine Sci. Gulf Caribb.* **8:** 278.

Tavolga, W. N., 1958*b*, The significance of underwater sounds produced by males of the gobiid fish *Bathygobius soporator*, *Physiol. Zool.* **31:** 259.

Tavolga, W. N., 1964, Sonic characteristics and mechanisms in marine fishes, *in* "Marine Bio-Acoustics" (W. N. Tavolga, ed.) pp. 195–211, Pergamon Press, Oxford.

Tavolga, W. N., 1965, Review of Marine Bio-acoustics. State of the Art: 1964, U.S. Naval training Device Center Technical Report 1212–1, Port Washington, N. Y.

Thorpe, W. H., 1961, "Bird Song," Cambridge University Press, Cambridge.

Thorpe, W. H., 1963, Antiphonal singing in birds as evidence for avian auditory reaction time, *Nature* **197:** 774.

Thorpe, W. H., and North, M. E. W., 1965, Origin and significance of the power of vocal imitation: With special reference to the antiphonal singing of birds, *Nature* **208:** 219.

Winn, H. E., 1967, Vocal facilitation and the biological significance of fish sounds, *in* "Marine Bio-Acoustics" (W. N. Tavolga, ed.) Vol. 2, pp. 213–230, Pergamon Press, Oxford.

Winn, H. E., and Stout, J. F., 1960, Sound production by the satinfin shiner, *Notropis analostanus*, and related fishes, *Science* **132:** 222.

Chapter 12

USING SOUND TO INFLUENCE THE BEHAVIOR OF FREE-RANGING MARINE ANIMALS

Arthur A. Myrberg, Jr.

School of Marine and Atmospheric Science
University of Miami
Miami, Florida

I. INTRODUCTION

The study of animal behavior holds many fascinations, but probably no aspect is more rewarding to its students than controlling, by appropriate stimuli, specific activities of free-ranging animals. The excitement that such work brings with it, especially when dealing with animals in the sea, stems in large part from the feeling that one has "tuned in" on an important line of communication with an animal on its terms—in an environment far different from that experienced by the researcher working under controlled laboratory conditions. It certainly is not my intent here to belittle laboratory studies; in fact, such studies often form the basis for work undertaken in the field. I am certain that much of our present knowledge about controlling the behavior of certain cetaceans and fishes rests, in large part, on the many hours of observation and experimentation that have been carried out on captive animals, held either in aquaria or in large enclosures. This will, I hope, become evident later in this chapter. Yet such results are not ends in themselves, although they may appear so at times. The control of overt behavior under field conditions will often serve to demonstrate that a hypothesis—set forth by preliminary observations, correlative analyses, and models of predictability—can increase our understanding of the activities of a given species in its natural environment, be it a coral reef, a grass flat, or a wide expanse of open sea.

Control of various behavioral activities may also occur, however, through trial-and-error practices such as presenting unrelated or unnatural stimuli to free-ranging marine animals. Such practices, though perhaps not adding immediately to our understanding of underlying behavioral mechanisms, may provide insight into the types of environmental information available to given species and provide the incentive for more applied research if the species is of economic value.

II. MAMMALS

There is no better way of introducing my subject than by recognizing the remarkably successful efforts of K. S. Norris, F. G. Wood, Jr., W. E. Evans, and their collaborators in controlling specific behavior of untethered dolphins (*Tursiops truncatus and Lagenorhynchus obliquidens*) and pinnipeds (*Zalophus californianus* and *Phoca vitulina*) in the open sea (Evans and Harmon, 1968; Norris, 1965; Ridgway, 1966; Wood and Ridgway, 1967). Their work with "Keiki," "Tuffy," and "Buzz-Buzz" has demonstrated the role that dolphins (and no doubt, in the near future, porpoises, seals, and perhaps even whales) may play in man's exploration of the sea.

An important technique used in their research has been to signal their subjects by appropriate acoustic "commands," associated through training procedures with acts leading to positive reinforcement. These "commands" have been (1) constant tones (cue to leave a cage and swim a measured course at high speed—Norris, 1965), (2) buzzers that produce a pulse train of intense broad-brand clicks (cue to swim to an underwater speaker and to stop directly in front of it—Norris, 1965; or then push a speaker with the snout—Ridgway, 1966), (3) buzzers with slightly different peak frequencies (cues to move between divers on the bottom—Wood and Ridgway, 1967), and (4) a combination of broad-band clicks with various buzzers (cues to deliver packages, e.g., tools and mail, to divers at varying depths and then to return to a specific location at the surface—Wood and Ridgway, 1967). The use of sound in this connection is easily understood when one considers the extremely sensitive hearing of these animals (Johnson, 1967; Møhl, 1968) and their powers of locating a source of such frequencies (Headquarters, Pacific Missile Range, 1965). Although considerable training in pens and enclosures precedes work in the open sea, the free environment provides the chance to accumulate important physiological and behavioral data about these animals which otherwise might escape us. Data, such as obtained by Ridgway (1966) and Ridgway *et al.* (1969) for the bottlenose dolphin (*Tursiops truncatus*) on diving depth, duration of dives, respiratory rates after dives, and O_2 utilization during dives, add valuable information to our limited knowledge about the habits and physiological capacities of these marine mammals.

Few attempts have been made to control the behavior (vocal or otherwise) of free-ranging cetaceans by playing back their own sounds, and the few results have appeared equivocal or negative (Watkins and Schevill, 1968; Busnel and Dziedzic, 1968). Yet, Busnel and Dziedzic (1966) have recorded, at various times, response patterns indicative of flight by *Delphinus delphis* when apparent stress sounds of the species (produced during capture attempts, harpooning, etc.) were played back to free-ranging schools. More recent experiments by Cummings and Thompson (1971) have provided us important information on the apparent role played by natural sounds during interspecific interactions. These authors played back natural sequences of "screams" of the killer whale (*Orcinus orca*) in the vicinity of a large group of gray whales (*Eschrichtius robustus*) as the latter moved along their usual migration paths off Point Loma, California. Surfacing whales immediately swirled around and headed directly away from the sound source. Individuals, near kelp, fled into its heavy growth and remained there ("spying" out over the surface) until the sound was stopped. Others, seaward of the research vessel, moved toward the open sea. Playback of other signals (random noise or pure tones) having the same on–off times and sound source levels as the killer whale "screams" produced little or no avoidance. The authors also noted that the grays emitted few sounds during transmission of the killer whale sounds. Fish and Vania (1971) also obtained remarkable success in preventing movement of belugas (*Delphinapterus leucas*) into the Kvichak River, Alaska, by projecting near its mouth high-level "screams" and echolocation-type pulses of their predator, the killer whale. Analogous to the previous experiment with grays, the belugas emitted few sounds during playback periods. I might add here parenthetically that the "screams" of the killer whale are of interest to our group at Miami as to their effects on "blue-water" sharks. Results, to date, appear sufficiently promising to continue preliminary testing with these animals (the sounds of killer whales were kindly supplied from tapes sent to us by Dr. Cummings).

Several laboratories have also directed considerable attention to the auditory capabilities of the pinnipeds. One member of this group, the Weddell seal (*Leptonychotes weddellii*), was studied in the field by Watkins and Schevill (1968). These authors reported that when conspecific sounds of high fidelity were transmitted into the sea, individuals responded to such sounds as if another seal were the source. High-frequency trills (many rapid pulses) and chirps (isolated pulses), transmitted through a projector placed below the sea-ice, resulted in behavioral correlates such as (1) vocal responses from various seals in the vicinity (responses sometimes appearing to be mimicking the transmitted sound) and (2) rapid movements of seals to the sound projector and, in some cases, even biting the projector. These activities were then followed, however, by an apparent loss of interest during repetitive stimulation. Such loss of interest, i.e., apparent habituation of response, has accompanied many playback studies among diverse groups of animals when positive reinforcement did not accompany the playback.

III. FISHES

Controlling the movements of free-ranging fishes, either directly (by attracting, by "exciting," or by guiding flight responses) or indirectly (by providing floating or submerged cover), has been attempted with apparently reasonable success by fishermen of many nations since ancient times. Fortunately, many of these practices have been documented by the excellent reviews of Westenberg (1953), Busnel (1959), Moulton (1963), and Wolff (1966). In these accounts, it is noteworthy that many practices relied on the ability of the fishes to respond to certain underwater sounds, the majority being of a broad-band nature, possessing transient qualities and pulse irregularities.

Studies by various fisheries interests have attempted to control movements of fishes by particular acoustic techniques. Unfortunately, the majority has been unsuccessful. Possible reasons for lack of success are legion; many have been discussed by Moulton (1963) and others. These need not concern us at this moment, except perhaps to point out that the more an experimenter knows, firsthand, about the behavior and sonic environment of his subjects, the greater his chance of success in contributing significantly to our knowledge.

Another relevant query into the acoustic biology of marine fishes is that of the significance of biological sounds in the sea, specifically the role that these sounds play in the lives of those fishes that either produce them or, at least, react to them (for a general review, see Freytag, 1968). A standard procedure for gaining such information is to play back specific sounds into the environment and watch for specific activities during transmission. Since it is difficult, however, to see very far underwater, studies have been confined either to quite shallow water or just below the surface. Moulton (1956) obtained vocal responses by sea robins (*Prionotus*) to sonic patterns having a temporal structure similar to their own sounds. The same author (1960) also observed specific movements of fish schools (*Anchoviella*) away from a source of artificially generated sounds or sounds recorded from predators (*Caranx*), as well as apparent attraction of still another predator (*Sphyraena*) to sounds of its prey (*Caranx*). Tavolga (1958) noted increased general activity of frillfin gobies (*Bathygobius soporator*) within the apparent range of a source transmitting their own sound, as well as rapid directed movements of probably mature males to the vicinity of the source. Shiskova (1958) reported that broad-band sounds. when projected at high intensity, resulted in schools of gray mullet swimming deep and then dispersing. Winn (1967) obtained vocal responses by toadfish (*Opsanus tau*) to playback of their own sounds and apparent inhibiting effects of background noise on such responses. He also mentioned evidence of possible antiphony. Fish (1968, 1969), also studying the sounds of the toadfish, found that males on their breeding nests varied their calling rates depending on the temporal patterning of either their own "boatwhistle" sound or a tone pattern. Calling rates were suppressed by playing back continuous pure tones, and antiphony could not be established. Finally, Hashimoto and Maniwa (1967) recorded apparent flight responses of yellow-tail snappers, barracuda, and jack mackerel to sounds produced by dolphins (*Grampus*) and attraction of impounded yellow-tail snappers to a source transmitting sounds recorded during their feeding periods.

Steinberg *et al.* (1965), using underwater television at Bimini (Kronengold *et al.*, 1964), reported various responses by fishes during periods of sonic playback at a depth of about 20 m. Slippery dicks (*Halichoeres bivittatus*) rapidly swam under the sound projector or burrowed into the sand when sounds associated with the approach of blue runners (*Caranx fusus*) were played back in their vicinity. Transmission of another sound, produced by a school of margate (*Haemulon album*) under attack by a barracuda, resulted in startle responses and crowding together by margates at the site. The authors also mention that yellow-tail snappers (*Ocyurus chrysurus*) were invariably attracted to pulsed 20 Hz signals and that a single mutton snapper (*Lutjanus analis*) was consistently attracted to an instrument package when its noisy switch was actuated.

With this background, I would like to report a series of experiments carried out on one of the most ubiquitous fishes found in the waters off south Florida and the Bahamas, the bicolor damselfish, *Eupomacentrus partitus* (Fig. 1). This study demonstrated the significance that certain sounds have on specific behavioral activities of fishes, both in the laboratory and the field, as well as the control that an experimenter may exercise over such activities.

In attempting a detailed behavioral study on unrestricted fishes at a depth where they were abundant, the major problem was to extend our eyes and ears to that depth, often for reasonably long periods of time. This was done by using the Bimini Video-Acoustic Installation (Fig. 2). For information pertaining to the original installation, see Steinberg and Koczy (1964), Kronengold *et al.* (1964), and Myrberg *et al.* (1966); its present modifications and new site are described by Stevenson (1967) and Myrberg *et al.* (1969). An underwater television (UTV) operates at a depth of 20 m,

Fig. 1. The bicolor damselfish, *Eupomacentrus partitus*, male, length 80 mm (TL). (Photo by Ed Fisher.)

Fig. 2. Divers working on the underwater television housing of the Bimini Video-Acoustic Installation. The UTV site is located 1.5 km off the west coast of North Bimini, Bahamas, at a depth of 20 m. (Photo by Ed Fisher.)

1.5 km off the west coast of North Bimini, near the eastern edge of the Gulf Stream (Fig. 3). The underwater scene is monitored on a screen in a laboratory housed on the grounds of the Lerner Marine Laboratory, North Bimini. When desired, the scene and sounds (picked up by an associated hydrophone) are recorded on magnetic video and/or sound tape. The camera housing is located in a rocky area, and its small size (1.0 m high, 0.5 m diameter) along with its long period of submergence (approximately 3.5 years) resulted in its blending remarkably well into the landscape.

The television site was selected so that, during periods of daylight,

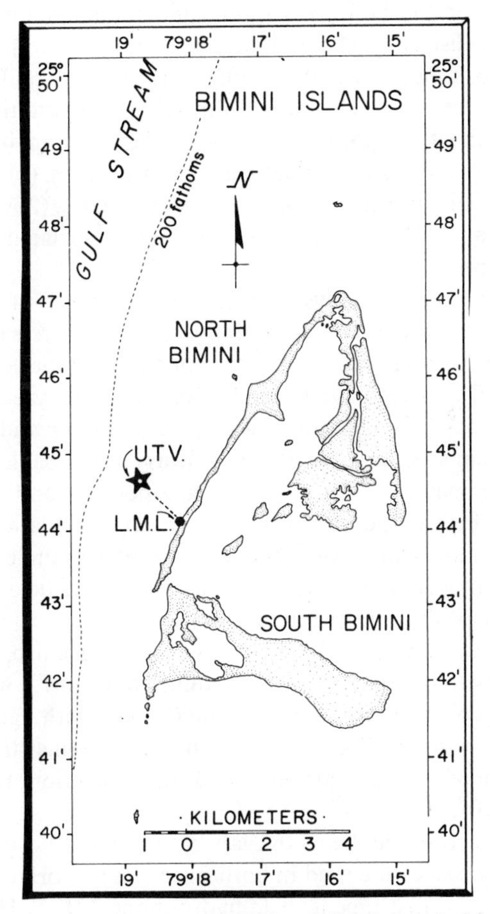

Fig. 3. The Bimini Islands, Bahamas. U.T.V.—underwater television site. L.M.L.—Lerner Marine Laboratory. Dotted line shows path of control cables. (From Myrberg *et al.*, 1969.)

observations and experiments could be undertaken on a small colony (three territorial males, six females, four juveniles) of bicolor damselfish. SCUBA was used for supplementary observations, specific measurements at the site, and preventive maintenance on the housing and associated underwater equipment.

Concurrent with the field study, observations were also conducted on a similar-sized colony of bicolors (three territorial males and five females, captured near the field colony) residing at the School of Marine and Atmospheric Science in an 800-liter tank, connected to an open seawater system (Wisby, 1964; Myrberg, 1969). The following report deals with these two specific colonies.

Our search for the biological significance of the sounds produced by the bicolor damselfish (Figs. 4 and 5) began, first, by spending over 6 months observing both colonies and making qualitative records and descriptions of their behavioral activities. This was followed by quantitative observations (Table I) on each of the three territorial males residing in the respective colonies. Their activity was emphasized because males (1) were extremely active in their social interactions compared with females, (2) were the primary sound sources, and (3) alone produced sounds associated with courtship patterns of the species.

Sequential contiguity analyses of 33 motor patterns (i.e., locomotory movements and postures) showed a nonrandom occurrence and temporal association into various groups of functional activities, e.g., courtship, agonism, nesting, and feeding. Records were also analyzed for the simultaneous occurrence of sonic patterns with specific motor and color patterns. Table II shows that specific sonic patterns clearly accompanied only a limited number of motor patterns of the male (see Appendix and Fig 6) and that both colonies reflected the same associations (the deviation in frequency of motor patterns associated with the "pop" sound was understandable since "chase" was the most common agonistic activity in the field, while "frontal-thrust" held that position in the laboratory).

Three sounds commonly heard ("chirp," "long chirp," and "grunt") were associated with courtship (a fourth sound, the "burr," was heard rarely and only during courtship in the field), one ("pop") with agonistic patterns, and one ("strid") with feeding. These specific associations strongly indicated that at least some of these sounds possessed signal function. Playback experiments were therefore initiated.

Sounds were recorded from the laboratory and field colonies (Crown International recorders were used in both instances; laboratory hydrophone-SB154B, Chesapeake Instrument; field hydrophone-2ZP15, Hudson Laboratories). Tape loops were then made. Each loop contained a single sound, and the length of the loop was adjusted so that the repetition rate of the sound was

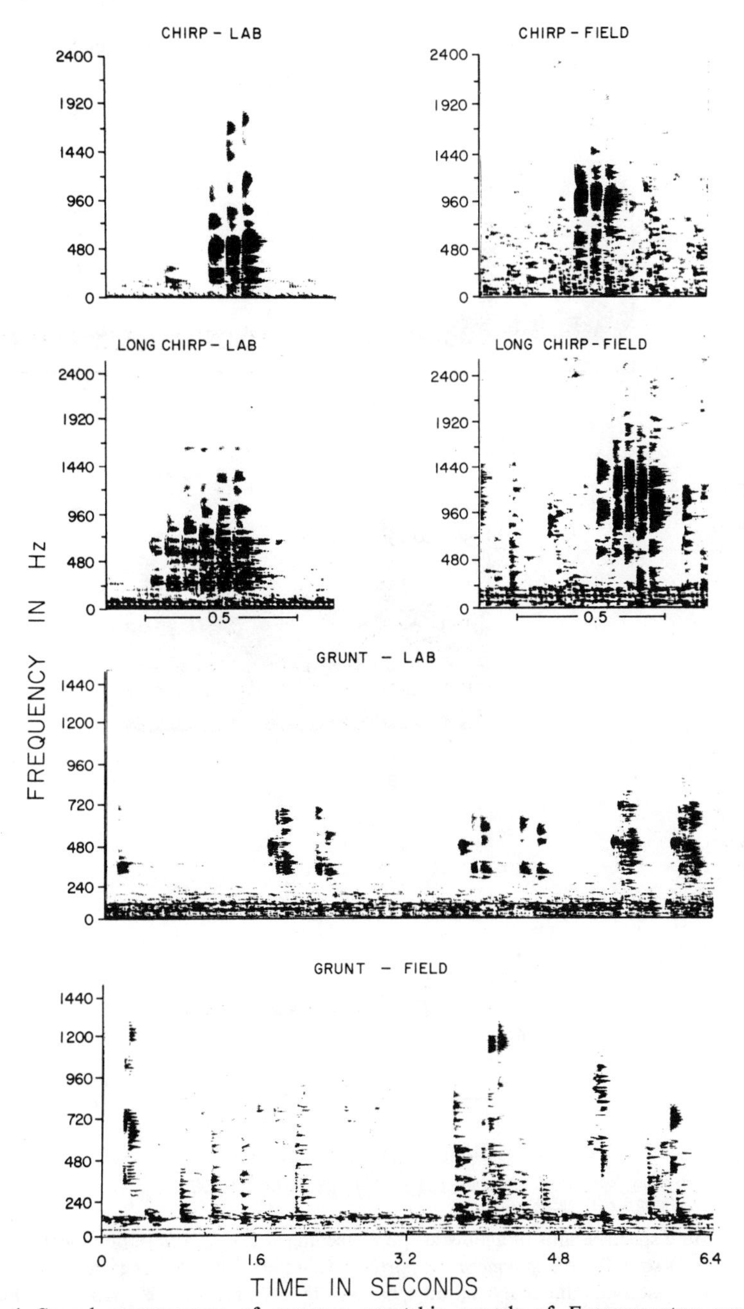

Fig. 4. Sound spectrograms of common courtship sounds of *Eupomacentrus partitus*. Comparative recordings from laboratory and field. The analyzing filter bandwidth is 20 Hz for the "chirp" and "long chirp" sounds and 6 Hz for the "grunt" sounds.

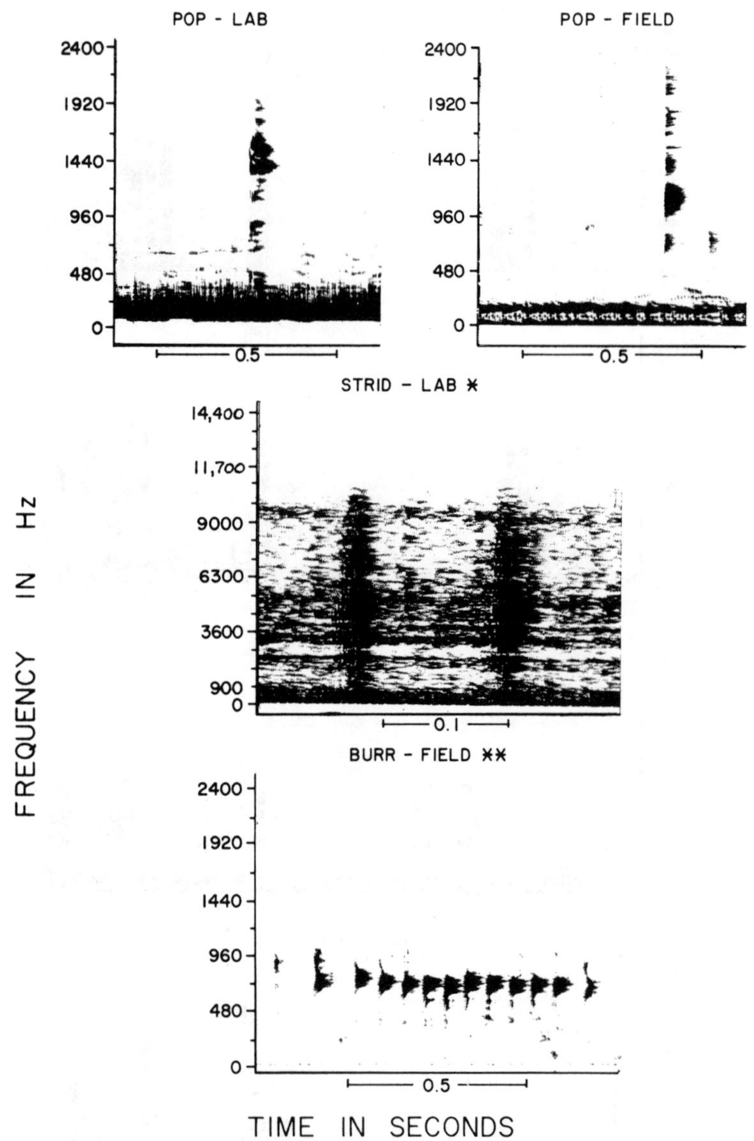

Fig. 5. Sound spectrograms of agonistic "pop", feeding ("strid"), and seldom heard courtship ("burr") sounds of *Eupomacentrus partitus*. Comparative recordings are given only for the "pop" sound. The analyzing filter bandwidth is 20 Hz for the "pop" and "burr" sounds and 60 Hz for the "strid" sounds.

Table I. Summary of Observation Periods During Which Data Were Obtained on the Simultaneous Occurrence of Motor, Sonic, and Color Patterns by Resident Males of the Laboratory and Field Colonies of Bicolor Damselfish (*Eupomacentrus partitus*)

Location of observation	No. periods 15 min	No. periods 30 min	Total No. minutes	Total No. action patterns recorded	Range of dates of observations
Laboratory	6	46	1470	28,382	17 Sept.–7 Oct., 1966
Field	48	16	1200	5,308	6 Sept.–2 Nov., 1966

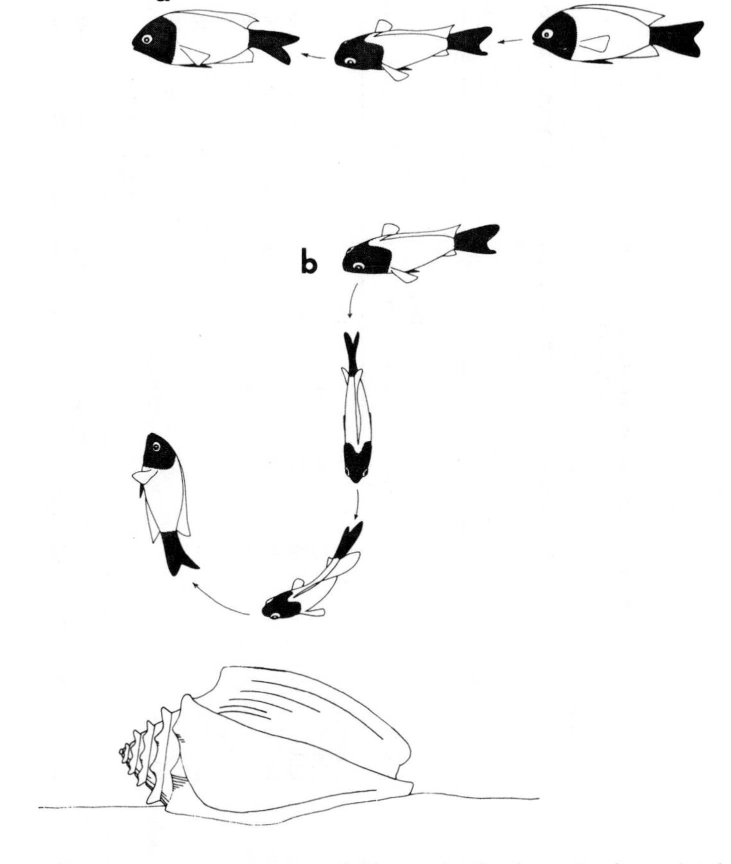

Fig. 6. Two common motor patterns (with associated color pattern) correlated with courtship in the bicolor damselfish. **(a)** "Tilt." **(b)** "Dip."

Table II. Simultaneous Occurrence of Specific Sonic Patterns with Specific Motor Patterns Shown by Male Bicolor Damselfish in Laboratory and Field Colonies

		No. times that a specific sonic pattern accompanied a specific motor pattern												
		Courtship patterns						Agonistic patterns						
Sonic pattern	Observations	Tilt	Dip	Lead	Flutter	Close-swim	Skim	Frontal-thrust	Chase	Circle	Half-circle	Tail-back	No. times sonic pattern occurred without motor pattern	Total occurrence of sonic pattern
Chirp	Laboratory[a]	1	1326	0	0	0	0	0	0	0	0	0	25	1352
	Field[b]	1	297	0	0	1	0	0	0	0	0	0	1	300
Long	Laboratory	0	0	2	41	0	0	0	0	0	0	0	0	43
Chirp	Field	0	0	0	14	0	0	0	0	0	0	0	0	14
Grunt	Laboratory	0	0	0	0	45	3	0	0	0	0	0	2	50
	Field	0	0	0	0	62	0	0	0	0	0	0	3	65
Pop	Laboratory	0	0	0	0	0	0	145	64	28	2	1	22	262
	Field	0	0	0	0	0	0	23	258	8	0	0	14	303
Stridulation	Laboratory	Associated with feeding on hard matter.												
	Field	Monitored only near hydrophone, associated with feeding on substrate.												

[a] Laboratory conditions: area 2.6 by 6 by 8 m; colony three territorial males, five females, no juveniles; 52 $\frac{1}{4}$–$\frac{1}{2}$ hr observations.
[b] Field conditions: area 4.5 by 4.0 by 1.3 m; colony three territorial males, six females, four juveniles; 64 $\frac{1}{4}$–$\frac{1}{2}$ hr observations.

similar to that produced by the species. Sounds were transmitted to the laboratory colony by a projector (J-9, Chesapeake Instrument) placed directly in its tank. Sound levels were determined during recording sessions by a VTVM, and recorded sounds were played back at that level. Laboratory tests were made during late fall (late November) because reproduction was at a low ebb. This meant that if males indeed responded to sonic playback, control periods (quiet periods of equal duration, bracketing each test) might reflect more accurately the difference in responsiveness since courtship patterns would then be effectively absent. Each test and control period was 5 min in duration. Approximately $\frac{1}{2}$ hr separated tests of different sounds. A day of "rest" separated each series. Three series were conducted (the specific test and control periods are shown in Figs. 8, 9, and 10). At least two independent observers were present during all testing.

Results of the laboratory study are summarized in Table III. All responses shown are from the three colony males; female activity showed no apparent change between test and control periods. Increase of motor patterns and sounds of courtship (the "chirp" was always associated with the "dip") during playback of courtship sounds and the lack of such activities during playback of the agonistic sound ("pop") demonstrated that sounds associated with courtship in the species clearly facilitated the initiation of courtship sequences, regardless of whether these sounds were recorded from the laboratory or field colony. The courtship sounds ("chirp" and "grunt") used in testing did not differ in their ability to bring forth courtship in males. Lack of responsiveness to the "pop" sound also demonstrated intraspecific sound discrimination by males of the species. Finally, the significant releasing effect of courtship sounds was evident when such sounds brought forth courtship during the reproductive ebb. There was no directed movement to the underwater projector by either sex during periods of transmission.

Figure 7 illustrates the three color patterns shown during the playback study. The uppermost (a) was the usual pattern shown by males as they fed, wandered about their range, and participated in low-level agonistic encounters (anterior portion of body and tail blue-black to gray; posterior portion of body light gray). The second pattern (b), termed the "white-body," was a paling of the posterior half of the body (except the tail). The third pattern (c), termed the "black-mask," was an extreme paling of the entire body, except for the head and tail, both of which became very dark (extreme paling to even a yellowish hue often extended along the nape to the region just above the eyes; this is not shown in the illustration). The temporal distribution of these color patterns during laboratory playback is seen in Figs. 8, 9, and 10. The lowermost pattern (d), termed the "white-bar," was a darkening of the entire body, except for a small area on the sides. This pattern was

Table III. Selective Responsiveness of Three Laboratory Males of *E. partitus* to Playback of Specific Sounds[a]

Sound type (source)	Recorded from	Trial situation	No. trials	Frequency				
				Specific motor patterns				Specific sonic pattern
				Tilt	Dip	Nudge	Lead	Chirp
Chirp *E. partitus*	Field	Control	10	0	0	0	0	0
		Test	10	56	54	3	5	33
	Laboratory	Control	5	0	0	0	0	0
		Test	5	34	34	2	5	23
Grunt *E. partitus*	Laboratory	Control	10	0	0	0	0	0
		Test	10	54	54	7	6	17
	Totals	Control	25	0	0	0	0	0
		Test	25	144	142	12	16	73
Pop *E. partitus*	Field	Control	4	0	0	0	0	0
		Test	4	0	0	0	0	0
	Laboratory	Control	9	1	0	0	0	0
		Test	9	0	1	0	0	0
	Totals	Control	13	1	0	0	0	0
		Test	13	0	1	0	0	0

[a] Comparisons are made between sounds recorded from the field and laboratory colonies. Sounds were played back at recording levels.

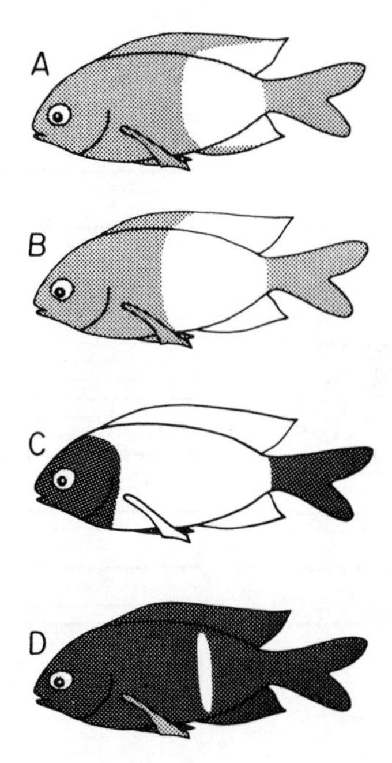

Fig. 7. Various color patterns of the male of *Eupomacentrus partitus*. (**a**) Normal pattern. (**b**) "White-body" (low-intensity courtship). (**c**) "Black-mask" (high-intensity courtship). (**d**) "White-bar" (high-intensity aggression).

shown by equally strong combatants during intense agonistic encounters, in contrast to the "black-mask" and "white-body," which were shown primarily during courtship activity only by males of both colonies. The "white-bar" was not observed during the playback series.

The "black-mask" and the "white-body" were clearly associated with the playback of courtship sounds. These two patterns appeared in control periods primarily during the first minute or so that followed a test of a courtship sound. Chromatic response to playback varied, but generally changes began between 30 sec to 1 min after onset of transmission. Courtship sounds, originating from the field, appeared initially to result in a slower arrival at the "black-mask" stage than when the subjects' own laboratory sounds were transmitted. The few tests made, however, preclude any stronger statement. The distribution of color patterns during periods of courtship and agonistic sounds demonstrated again the close tie that exists between certain sonic, motor, and color patterns in the species. I might add that, although not indicated in the illustrations, males appeared to show a decrement in motor response as testing proceeded. This apparent habitua-

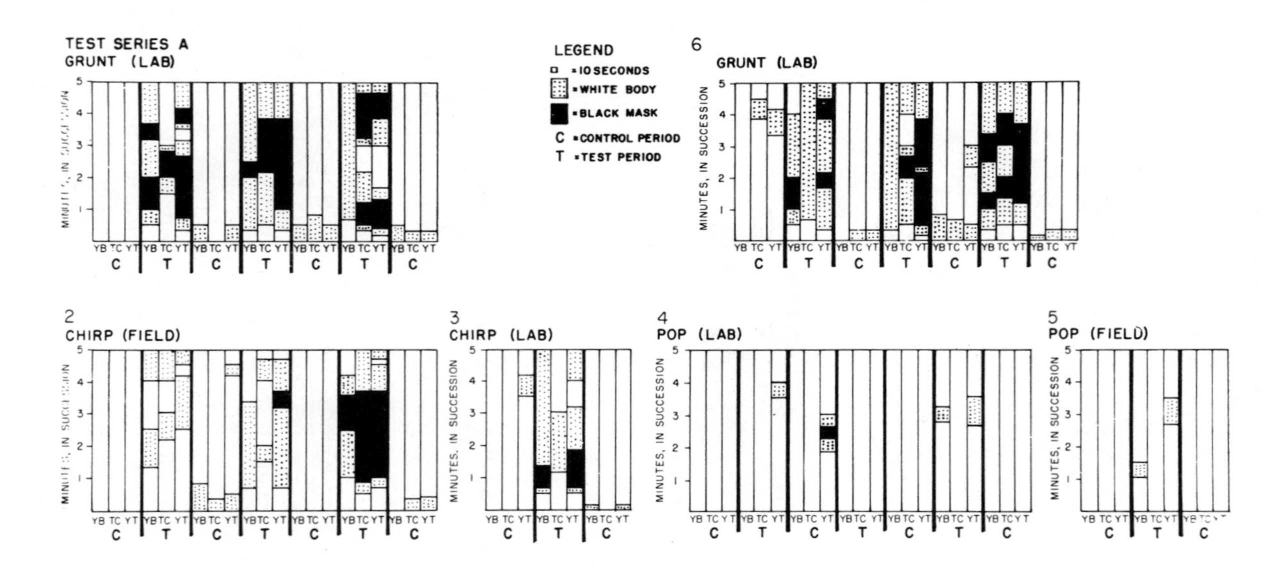

Fig. 8. Temporal changes in color patterns of each of three test males (YB, TC, YT) from the laboratory colony during playback of intraspecific sounds—test series A. Absence of "white-body" or "black-mask" (courtship patterns) during series implies presence of normal (noncourtship) color pattern. Test (T) and control (C) periods were each 5 min in duration. Further information is supplied in text.

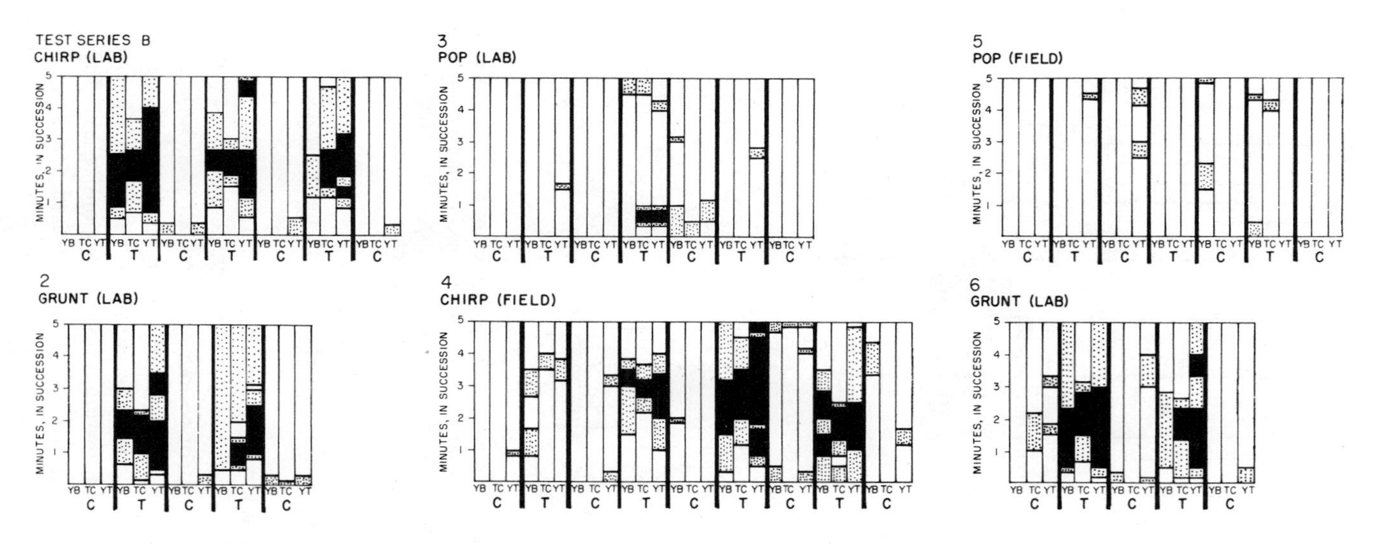

Fig. 9. Temporal changes in color patterns of each of three test males (YB, TC, YT) from the laboratory colony during playback of intraspecific sounds—test series B. Absence of "white-body" or "black-mask" (courtship patterns) during series implies presence of normal (noncourtship) color pattern. Further information is supplied in Fig. 8 and text.

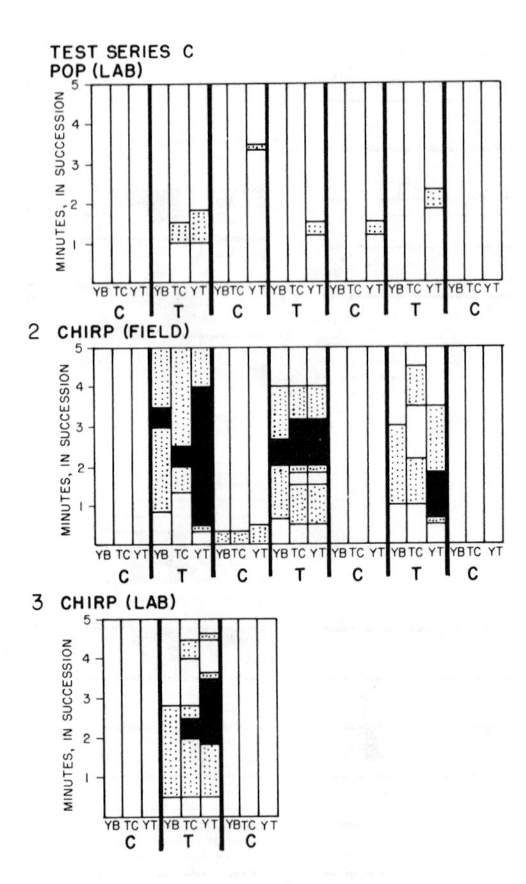

Fig. 10. Temporal changes in color patterns of each of three test males (YB, TC, YT) from the laboratory colony during playback of intraspecific sounds—test series C. Absence of "white-body" or "black-mask" (courtship patterns) during series implies presence of normal (noncourtship) color pattern. Further information is supplied in Fig. 8 and text.

tion to the stimulus situation has shown itself more strongly in more recent testing.

The field colony was next studied, but not until early spring. Courtship was occurring regularly at that time, and therefore associated activities were present during control periods. This actually was fortunate since a low level of activity during those periods might result in a decrease, as well as an increase, in response toward a particular sound pattern. Such a decrease might therefore indicate an inhibition of responsiveness. Tape loops used previously during the laboratory tests were used again for the field tests. Sounds were transmitted through a projector (J-9) located a few meters from the field colony. As in the case of the laboratory tests, at least two independent observers were present during all field tests.

Table IV summarizes the selective responsiveness of males of the field colony to playback of courtship and agonistic sounds recorded from both the laboratory and field colonies, as well as the "staccato" call of the squirrelfish, *Holocentrus rufus* (Winn and Marshall, 1963; Winn *et al.*, 1964). The latter species was commonly found on outcrops surrounding the UTV site. There was a significant increase in the occurrence of specific motor and sonic patterns of courtship during transmission of courtship sounds over that shown during control periods. These results mirrored those obtained previously with the laboratory colony. The conclusions based on the laboratory testing again held true—clear discrimination between sounds and clear facilitating effects of courtship sounds on the initiation of courtship in males. Additionally, the significant decrease of courtship during playback periods of the "staccato" and "pop" sounds indicated that such sounds can inhibit, for at least short periods, courtship in the species. Table V summarizes all testing and illustrates the similarity between results obtained on both colonies—with the single exception that absence of courtship patterns in the laboratory during control periods precluded the possibility of demonstrating the inhibiting effect noted in the field. Color changes were not quantified during field testing because of difficulty in determining precisely the onset of the "white-body" and the "black-mask." Qualitative records, however, paralleled those from the laboratory. Apparent habituation to playback was again evident in the field tests, and it appeared to occur more rapidly than that noted in the laboratory. Another playback series was begun in late spring for purposes of determining whether we could increase spawning activity. The very high level of courtship activity during periods of no transmission, however, prevented further testing at that time. Sound playback experiments, using the clear responses of bicolors to their various sounds, have fortunately been continued by one of my students, Samuel J. Ha. His interest lies primarily in determining those parameters of the various sounds which provide the information necessary to bring forth specific responses.

The investigations just described have shown clearly that one can control by appropriate sonic playback either an increase or a decrease in specific sonic, motor, and color patterns in either restricted or unrestricted fish of at least one species. This, I believe, is an important finding, having many ramifications. Of course, bicolor damselfish are small, ubiquitous reef fishes and their control is of importance probably only to behavioral scientists. Some additional work that we have carried out along the lines of influencing the movements of larger, free-ranging fishes will perhaps be of broader interest.

Table IV. Selective Responsiveness of Three Field Males of *E. partitus* to Playback of Specific Sounds[a]

Sound type (source)	Recorded from	Trial situation	No. trials	Frequency				
				Specific motor patterns				Specific sonic pattern
				Tilt	Dip	Nudge	Lead	Chirp
Chirp	Field	Control	7	12	10	0	0	8
E. partitus		Test	7	71	71	3	5	55
	Laboratory	Control	4	7	7	1	0	7
		Test	4	29	29	0	3	26
Grunt	Laboratory	Control	8	8	8	0	0	6
E. partitus		Test	8	64	64	6	8	56
	Totals	Control	19	27	25	1	0	21
		Test	19	164	164	9	16	137
Pop	Field	Control	8	13	12	0	1	10
E. partitus		Test	8	9	8	0	0	7
	Laboratory	Control	7	14	13	0	0	12
		Test	7	8	8	0	0	7
Staccato	Field	Control	4	6	5	0	0	5
Holocentrus rufus		Test	4	1	1	0	0	1
	Totals	Control	19	33	30	0	1	27
		Test	19	18	17	0	0	15

[a] Comparisons are made between sounds recorded from the field and laboratory colonies. Sounds were played back at recording levels.

Table V. Summary of Tests, Demonstrating Selective Responsiveness of Males of *E. partitus* to Playback of Specific Sounds[a]

Location of playback	Sound types	Trial situation	No. trials	Frequency				
				Specific motor patterns				Specific sonic pattern
				Tilt	Dip	Nudge	Lead	Chirp
Laboratory	Courtship	Control	25	0	0	0	0	0
	(chirp, grunt)	Test	25	144	142	12	16	73
Field	Courtship	Control	19	27	25	1	0	21
	(chirp, grunt)	Test	19	164	164	9	16	137
Laboratory	Noncourtship	Control	13	1	0	0	0	0
	(pop)	Test	13	0	1	0	0	0
Field	Noncourtship	Control	19	33	30	0	1	27
	(pop, staccato)	Test	19	18	17	0	0	15

[a] Same sound tapes used in field and laboratory, three males observed during each trial (duration 5 min).

For the last few years, we have been interested in the hearing abilities of sharks. Although considerable data have been obtained on laboratory subjects (Nelson, 1965; Banner, 1967*a,b*), attempts have also been made to determine the importance of underwater sounds to free-ranging sharks. Since completion of the first study along these lines (Nelson and Gruber, 1963—evidence that certain species of free-ranging sharks were attracted to various acoustic signals), this interest has continued at our institute (e.g., Banner, 1968, 1970; Myrberg, 1968, in press) and elsewhere (e.g., Nelson; 1968, Nelson and Johnson, 1970*a,b*; Nelson *et al.*, 1969). Concurrent with this research, other investigators have attempted to attract free-ranging sharks to sources of underwater sounds, and their varying results have been discussed at recent meetings. Since various published reports also indicated that certain classes of sounds were ineffective in attracting sharks (Hobson, 1963; Wisby *et al.*, 1964), we thought that a detailed investigation on this question was in order.

Using, once again, the Bimini Video-Acoustic Installation, this investigation was completed (Myrberg *et al.*, 1969). It dealt with (1) differences in the attractive nature of various types of acoustic signals, (2) identification of those sharks that arrived at the sound source during periods of transmission, (3) apparent habituation to acoustic signals in the absence of apparent positive reinforcement, (4) upper frequency limit for purposes of attracting sharks in the area, and (5) demonstration of hearing, as well as indication of directional orientation, by sharks in the acoustic far-field.

Sounds transmitted by a J-9 projector, as well as ambient noise, were received by a calibrated pressure hydrophone positioned 18.5 m from the projector. A VTVM and tape recorder (Uher-4000 Report L) were used for monitoring, while a variable bandpass filter was used for determining signal-to-noise relationships and to control frequency cut-off of selected, random-noise signals. Sine wave signals passed directly to a 50-W amplifier, used to drive the projector. Either by overdriving that amplifier or by using it within its normal range, two classes of signals were produced—pure tones and over-driven sine waves. The latter resulted in odd-harmonic, biphasic square waves. Broad-band sounds formed a third class. These were produced with a white-noise generator and passed successively through the bandpass filter and a photoswitch (used to eliminate transients) before entering the driving amplifier. Signals were also monitored at the amplifier output by an oscilloscope to assure the quality of a given signal at that point in the system. Sound levels were determined shortly before testing by projecting a brief, continuous burst of sound. Most signals were maintained slightly more than 20 db above the level of broad-band ambient at the hydrophone. All signals were pulsed rapidly and irregularly by manually keying the system.

A test consisted of playing a given signal for 3 min. A 1-min rest period separated each test from that of a control period (quiet periods of equal duration). This "rest" allowed the control period to reflect more accurately the differences in responsiveness between times when signals were transmitted and times when they were not. Records were kept of (1) the duration (in seconds) between the onset of a given period and the arrival of the first shark, (2) the total sightings for a given period (based on the frequency of sharks passing through the field of view), and (3) the maximum number of sharks seen at a single time on the monitor screen for each period.

Table VI summarizes the results obtained during the investigation. It is obvious that certain acoustic signals, such as overdriven sine waves and broad-band filtered sounds—in the frequency range tested—attracted sharks to the area of the sound source. Any sound, however, will not necessarily attract sharks, as demonstrated by tests using pure tones. It is important to note that only some tens of seconds or a few minutes separated onset of the signal and appearance of the first shark at the site. Although it certainly is possible that sharks may have remained in the immediate area once having been attracted, such a possibility was precluded during the first test of a given day. Divers reported that no sharks were evident within a 50 m radius of the UTV immediately prior to that test.

Most sharks attracted to the site were sharpnose sharks, *Rhizoprionodon* sp. Two reef sharks, *Carcharhinus springeri*, were also identified as well as one shark identified either as a silky shark, *C. falciformis*, or a dusky shark, *C. obscurus*, and a nurse shark, *Ginglymostoma cirratum*.

Figure 11, showing results from three experimental series, illustrates the point that sharks approached the site only during the test periods (in this case, signals were irregularly pulsed, overdriven sine waves). The entire series, covering different times of the day—morning, noon, and late afternoon—also demonstrated approach by sharks to a sound source at any time of the day. Trends indicated decreasing numbers of sharks at the site during successive testing, and apparent habituation was therefore checked. Figure 12 shows that such a phenomenon was apparently playing an important role in our testing. It appeared that our sharks were "learning to ignore the signal." Minimum time required to regain a prehabituation level of sightings was not investigated, but within about an hour after the last test of that series that level was again achieved. Still other tests provided us an indication of the minimum effective signal for purposes of attraction at the site. Broad-band filtered noise, 500–1000 Hz, was the signal of interest, as it had attracted a goodly number of sharks. Spectrum level measurements of this signal showed it to be about 20 db above spectrum level ambient at the same frequency. The broad-band level was approximately 3 db above

Table VI. Acoustic Attraction of Sharks: Summary of Experimental Series[a]

Acoustic signal (irregularly pulsed)	Breakdown of test periods: Number of periods	Total shark sightings	Maximum simultaneous shark sightings per period	Time between onset of signal and first shark sighting (sec)
Overdriven sine waves[b]				
55	5	37	4	28–46
70	2	12	2	30–51
80	9	83	6	23–76
100	2	16	2	22–31
120	1	2	1	125
200	3	51	5	27–31
500	3	117	8	11–54
800	2	1	1	141
1000	2	0	0	—
1200	2	0	0	—
1500	2	0	0	—
	Total 33	319		
Broad-band filtered				
25–50	2	2	1	110
50–100	2	5	2	.45
150–300	2	2	1	122
400–800	2	3	1	21
500–1000	9	44	5	14–32
	Total 17	56		
Pure sine waves				
55	1	0	0	—
80	4	0	0	—
120	1	0	0	—
200	2	0	0	—
400	2	0	0	—
500	5	0	0	—
600	2	0	0	—
700	2	0	0	—
800	2	0	0	—
1000	1	0	0	—
1200	1	0	0	—
1500	1	0	0	—
	Total 24	0		

[a] From Myrberg *et al.* (1969); 87 control periods with six sharks sighted and 74 test periods with 375 sharks sighted.
[b] Alternating, symmetrical, and distorted square waves including harmonics.

broad-band ambient. This minimum effective signal (surely not a meaningful threshold determination) was, perhaps coincidentally, approximately that found in threshold determinations on young lemon sharks (*Negaprion brevirostris*) tested in the laboratory at similar frequencies (Banner, 1967*b*). Unfortunately, signal-to-noise ratios from field and laboratory experiments

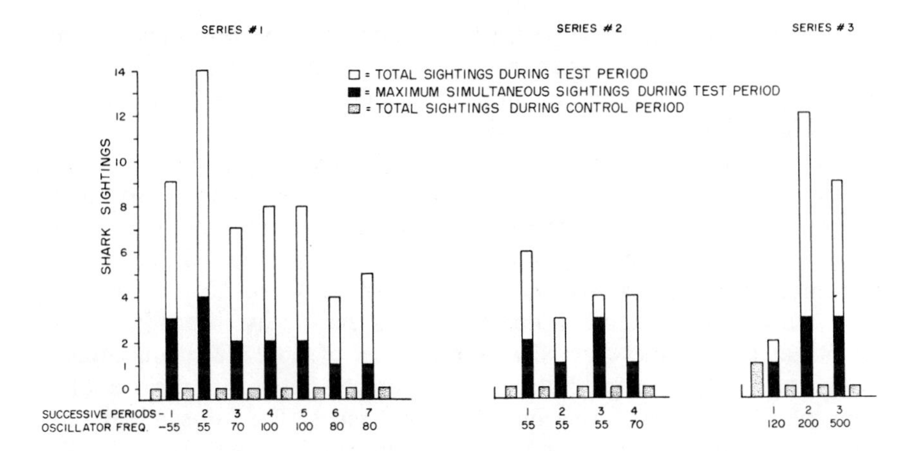

Fig. 11. Attraction of sharks by acoustic signals. Signals consisted of irregularly pulsed, overdriven sine waves (biphasic, symmetrical, and distorted square waves). Peak sound pressure level at 18.5 m from sound source was approximately 20 db above broad-band ambient noise. Each test and control period 3 min. Start of series 1 at 1650 hr, series 2 at 0920 hr, series 3 at 1300 hr. (From Myrberg *et al.*, 1969.)

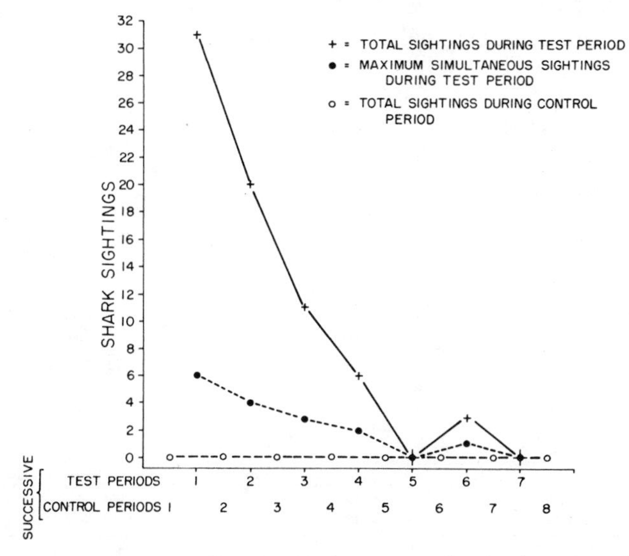

Fig. 12. Attraction of sharks by acoustic signals. Decrease in sightings through successive test periods. Signals consisted of constant-level, irregularly pulsed, overdriven 80 Hz sine waves (biphasic, symmetrical, and distorted square waves). Peak sound pressure level at 18.5 m from sound source was approximately 20 db above broad-band ambient noise. Each test and control period 3 min. (From Myrberg *et al.*, 1969.)

are not directly comparable. Nevertheless, attraction by weak signals (peak frequency of the signal was about −24 dbμb at the hydrophone) indicates the importance of acoustic stimuli to these animals. For further details, see Myrberg *et al.* (1969).

During the last-mentioned investigation, various large teleosts were also attracted to the vicinity of the sound source. The activity of sharks during testing precluded, however, exact records from being kept on these fishes. The results, nevertheless, agreed well with those obtained by Joseph Richard (1968), a member of our group, who, also using the Bimini Video-Acoustic Installation, attracted teleosts to a source of low-frequency octave bands (primarily 25–50 Hz) at intensities that fell within the range of those used in our investigation on sharks. Figure 13 illustrates the order of arrival of various fishes into the field of view during one of his test series. The fish symbols used in this illustration are positioned vertically in an arbitrary manner, and the specific groupings in no way imply schooling structure. Schools of margate (*Haemulon*) and yellow-tail snapper (*Ocyurus*) were observed during other tests, but these are not shown in the figure. Figure 14 summarizes the results of the entire investigation carried out by Richard, showing the relative attraction of sharks and teleosts.

It is apparent from the above data as well as from those obtained from others (e.g., Nelson and Johnson, 1970*a*; Nelson *et al.*, 1969; Wisby *et al.*, 1964) that low-frequency, pulsed, acoustic signals are an effective attractant

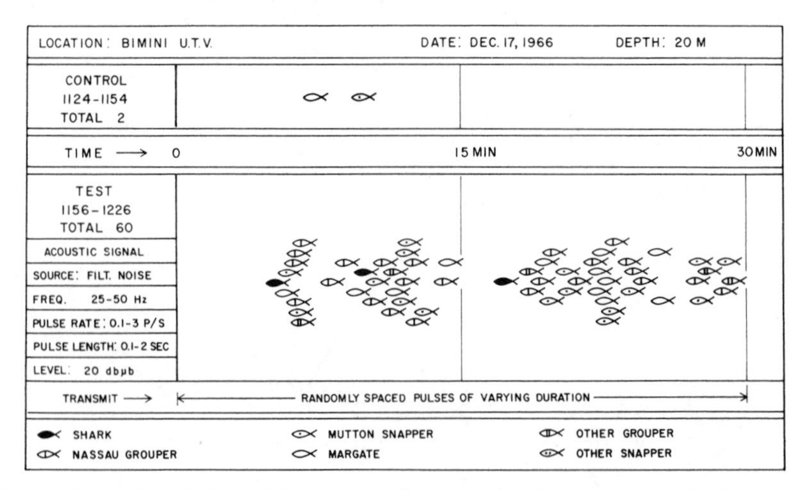

Fig. 13. Attraction of teleost fishes to the underwater television site by acoustic signals. One experimental series. Test and control periods are shown. Acoustic signals: randomly pulsed noise, 25–50 Hz. (Reprinted with permission from *J. Fish. Res. Bd. Canada* **25(7)**:1441–1452, Richard, 1968.)

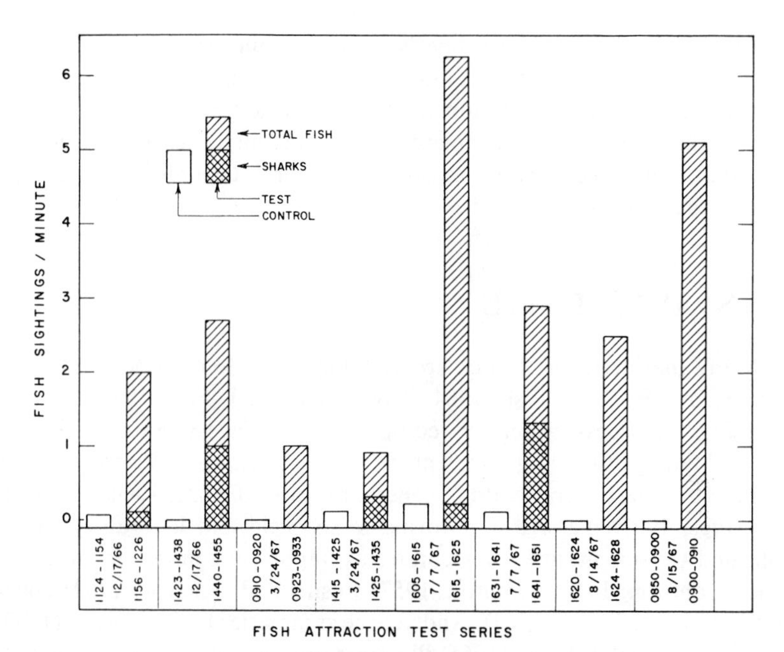

Fig. 14. Summary of test results showing the relative attraction of sharks and teleosts. Acoustic signals in the majority of tests were randomly pulsed noise, 25–50 Hz, and in others were randomly pulsed noise, 100–200 Hz and broad-band. (Reprinted with permission from *J. Fish. Res. Bd. Canada* **25(7)**: 1441–1452, Richard, 1968.)

for at least some predatory fishes. Results strongly indicate, however, that these same animals are not attracted by pulsed, pure tones (within frequencies from 50 to 1500 Hz) or continuous noise signals. Apparently, substantial bandwidth and continuous amplitude modulation (pulsing) are essential characteristics of an attractive acoustic signal. Unfortunately, results of these investigations reveal little about the relative effectiveness of different acoustic signals. An investigation is presently underway that will delineate, if possible, the frequency spectra and pulse repetition characteristics which are most effective for attracting various species of "blue-water" sharks, e.g., the silky shark (*Carcharhinus falciformis*). Members of this species have frequently been attracted to our sound source during ocean testing, as well as to a source used by Nelson *et al.* (1969).

Marine biologists are well aware of the inadequacy of our knowledge concerning relationships between various environmental factors and the activities of marine animals. As our knowledge of these relationships increases, so will our ability to predict, and even control in some cases, the behavior of these animals. Technical systems, such as the Bimini Video-

Acoustic Installation—when properly conceived and used—will aid immeasurably in this effort. The advantages of this system for purposes of field observation and experimentation have provided a clear idea of the important role that the sonic environment has in directing at least some activities of sharks, large predatory fishes, and even our little ubiquitous friend of the reef, the bicolor damselfish.

IV. INVERTEBRATES

Many marine invertebrates are certainly capable of producing sounds (Dumortier, 1963*a*; M. Fish, 1964; Tavolga, 1965; J. Fish, 1966) as well as possessing the necessary sensory equipment to receive vibratory stimuli of even slight magnitude (e.g., coelenterates—Josephson, 1961; crustaceans—Green, 1962, and Laverack, 1962; general reviews—Frings, 1964, and Frings and Frings, 1967). Passive monitoring in the field and laboratory has also implicated various marine invertebrates as important sources of underwater sounds (Johnson, 1943; Moulton, 1957; Busnel and Dziedzic, 1962; Hazlett and Winn, 1962; J. Fish, 1964). And yet the summaries by Dumortier (1963*b*), Frings (1964), and Tavolga (1965) illustrate our lack of knowledge regarding the biological significance (or the possible control) of sounds produced by marine invertebrates. Difficulties in observing the field behavior of small marine invertebrates constitute probably the largest obstacle to further knowledge. The numerous acoustic problems faced by biologists when studying the acoustic biology of underwater animals, particularly small invertebrates, have been reviewed by Frings (1964). Although problems are many, these may be reduced by the recent advances in diving techniques, underwater television, underwater manned habitats, small quiet submersibles, and underwater acoustic instrumentation.

The excellent studies carried out by Salmon and his colleagues on the semiterrestrial fiddler crab (*Uca*) demonstrate the exciting research that can be undertaken with marine invertebrates. Their work (Salmon and Stout, 1962; Salmon, 1965, 1967; Salmon and Atsaides, 1968) has clearly shown the use of sound in the reproductive behavior of *Uca pugilator*. Playing back specific sounds produced by males, for example, has resulted in (1) initiating sound production by other males near the source, (2) increasing the repetition rates by males already producing sounds, (3) inducing males in their burrows to come to the surface, and (4) increasing repetition rates of "waving" the large chela (a specific movement associated with courtship in the species). I am certain that in the near future additionally exciting studies will be carried out along similar lines with other invertebrates.

The acoustic behavior of unrestricted spiny lobsters and the specific control of movements by these animals have been, and continue to be, of interest to various investigators. Lindberg (1965) observed that when stridulating surfaces of *Panulirus interruptus* were rubbed together in the presence of unrestricted lobsters, the latter became alert and backed away into their lairs. Such activity occurred invariably when stridulation (rasp ?) occurred within 2 ft of a sheltered lobster, regardless of whether or not the stridulating lobster could be seen. Joseph Richard's interest in the behavior of another spiny lobster, *Panulirus argus*, has resulted in laboratory observations on the responses of these animals to simulations of their own stridulation sounds, and field studies on unrestricted individuals are planned in the near future (Richard, personal communication).

APPENDIX

An inventory (ethogram) of all recognizable species-typical motor patterns of *E. partitus* (laboratory and field) has been completed (Myrberg, in press). The following is a brief description of the motor patterns mentioned in this chapter.

The "tilt" (Fig. 6a) is a lateral tilting of the body, 20–45°, followed by a return to the normal vertical position. It can occur while a male remains motionless, but it is most often seen when the fish is moving about the water column. The "tilt" is often followed by the "dip" (Fig. 6b), which is a vertical or near-vertical dive by a male toward the substrate from a position 1–2 m above the substrate. After dipping a number of times, the male pitches his body slightly, head down (pelvics slightly expanded), and then, holding that position, rapidly swims toward a female. This pattern, termed the "nudge," is directed toward the ventral side of the caudal peduncle of the female. Often contact is not made, however, since the unique mode of swimming does not allow reaching a rapidly moving female. Following the "nudge," the male often shows the "lead," which is a rapid movement to the nest area. This pattern, showing exaggerated tail movements, occurs as the female follows the male into his territory. Upon entering the nest area, the male then performs the "quiver," a rapid quivering of the body while "standing on his head" directly at the prospective spawning site. As the female nears the nest, the male moves toward her and both swim about an area, 10–15 cm in diameter, with bodies almost touching. This pattern, the "close-swim," is also characterized by the two fishes often swimming in opposite directions a few centimeters above the substrate with the dorsal fins usually fully expanded. After "close-swim" has proceeded for 10–15 sec, the male moves rapidly to the spawning site and proceeds to "skim." This distinctive pattern is characterized by the male moving his venter just above the substrate (wall,

Sargassum, conch, etc.). The body, vibrating slightly, moves forward for a distance of 10–15 cm. The head is held up a bit higher than the rest of the body, and the latter takes on the slightest arch. If, during "quiver," "close-swim," or "skim," the female suddenly moves out of the nest area, the male then performs the "flutter." He moves to just above the substrate, tilts his body 20–45° to the side, and, with all fins fully spread, swims rapidly toward the female for some distance—up to 1 m—and then swims backward a short distance, followed by, again, a short forward movement. These rapid forward and backward movements can occur three to six times in a matter of 5–7 sec. The male may, at times, tilt sideways almost 90° from the vertical during such rapid movements. The sequencing of motor patterns described is not rigidly fixed. Rather, the patterns follow one another in a statistical sense: given that a particular pattern has occurred, there is a higher probability of a second pattern occurring next rather than any other pattern. These actions are considered courtship patterns on the basis of quantitative analyses to be covered elsewhere (Myrberg, in press), but it is also sufficient to characterize them as courtship activities because, although they often occur without spawning, they always precede it, when such occurs. Finally, these patterns are all closely correlated with the color pattern "black-mask" (Fig. 7c).

The second group of motor patterns mentioned in this chapter deals with but a few of the numerous agonistic patterns exhibited by the species. These patterns are closely associated contiguously with those patterns that characterize agonism in this species as well as in other species, i.e., chasing, biting, and fleeing. They are often characterized (except when brief, rapid fights occur) also by a color pattern, the "white-bar" (Fig. 7d). The "frontal-thrust" is a rapid forward movement toward a nearby adversary. The unpaired fins are expanded while the pelvic fins are held tightly against the body. The body is pitched slightly, head upward; the mouth is slightly opened. The "chase" is a rapid swimming toward a fleeing individual; the fins are held close to the body. Maximum sustained speeds are attained during the "chase," as well as during its counterpart, "flight." "Half-circle" is a sudden turning by a fish in "flight." It suddenly checks its speed and veers around to face, head on, the chaser. This is often followed by a "frontal-thrust," a "frontal-display" (same as "frontal-thrust," but no movement), or "flight," once again. The pattern, termed the "circle," is movement about a small, imaginary circle by two individuals with heads often opposed or head to tail. Fins are fully expanded, but slight pelvic fin and tail movements occur, as circling varies in speed. Finally, "tail-back" occurs in males with fully expanded fins and a backing-up of the body, tail first, toward an adversary. The expanded tail sometimes sweeps rapidly back and forth, giving the impression of a tail-beat (often seen in cichlid fishes, but there the orientation is far different). In females and low-ranking males, the fish backs up with lowered fins and no tail-beating is evident.

ACKNOWLEDGMENTS

The development and operation of the Bimini Video-Acoustic Installation as well as various biological programs (e.g., damselfish and shark research) undertaken with the Installation were supported by the Office of Naval Research, Oceanic Biology Programs, contracts Nonr 840(13), Nonr 4008(10), and N00014-67-A-0201-0004. Additional support was also given by the National Science Foundation, grants GB3234 and GB5897. The author wishes to thank C. Richard Robins and William C. Cummings for first directing his interest to the bicolor damselfish. Critical discussions with Robert A. Stevenson, Jr., Joseph D. Richard, John C. Steinberg, and Arnold Banner about test designs, technical equipment, and interpretation of results were indispensable for the progress of the research detailed here. The difficult preparation of observational data for computer analyses within the program dealing with the bicolor damselfish was handled almost exclusively by Leon T. Davies. His abiding interest and efforts in that regard are deeply appreciated. Thanks are directed also to Juanita Y. Spires, Samuel J. Ha, Charles Gordon, and Theodore Crabtree for their outstanding technical assistance in many phases of the research. Appreciation is directed to Stanley Walewski, Samuel H. Gruber, and Robert A. Stevenson, Jr., for their indispensable aid during diving operations. Robert F. Mathewson, Director of the Lerner Marine Laboratory, Bimini, and his staff provided the kindest hospitality and aid while various of these investigations were being carried out at the LML. This is contribution No. 1421 from the Rosenstiel School of Marine and Atmospheric Science, University of Miami.

REFERENCES

Banner, A., 1967*a*, Sensitivity to acoustic displacements in the lemon shark, *Negaprion brevirostris* (Poey), Masters thesis, University of Miami, 34 pp.

Banner, A., 1967*b*, Evidence of sensitivity to acoustic displacements in the lemon shark, *Negaprion brevirostris* (Poey), in "Lateral Line Detectors" (P. H. Cahn, ed.) pp. 265–273, Indiana University Press, Bloomington.

Banner, A., 1968, Attraction of young lemon sharks, *Negaprion brevirostris*, by sound, *Copeia* **1968(4)**: 871–872.

Banner, A., 1970, The use of sound in predation by young lemon sharks, *Negaprion brevirostris*, Doctoral dissertation, University of Miami, 100 pp.

Busnel, R.-G., 1959, Étude d'un appeau acoustique pour la pêche, utilisé au Sénégal et au Niger, *Bull. de l'IFAN* **21**, *Ser. A* (**1**): 346–360.

Busnel, R.-G., and Dziedzic, A., 1962, Rhytme du bruit de fond de la mer a proximité des côtes et relations avec l'activité acoustique des populations d'un cirripede fixe immergé, *Cahiers Oceanographique* **14(5):** 293–322.

Busnel, R.-G., and Dziedzic, A., 1966, Acoustic signals of the pilot whale *Globicephala melaena* and of the porpoises *Delphinus delphis* and *Phocoena phocoena*, *in* "Whales, Dolphins and Porpoises" (K. S. Norris, ed.) pp. 607–646, University Press, Berkeley.

Busnel, R.-G., and Dziedzic, A., 1968, Étude des signaux acoustiques associés a des situations de détresse chez certains cétacés odontocètes, *Ann. Inst. Océan.* **46:** 109–144.

Cummings, W. C., and Thompson, P. O., 1971, Gray whales, *Eschrichtius robustus*, avoid the underwater sounds of Killer whales, *Orcinus orca*, *Fish. Bull.* **69(3):** 525–530.

Dumortier, B., 1963*a*, Morphology of sound emission apparatus in Arthropoda, *in* "Acoustic Behavior of Animals" (R.-G. Busnel, ed.) pp. 277–345, Elsevier Publ. Co., New York.

Dumortier, B., 1963*b*, Ethological and physiological study of sound emissions in Arthropoda, *in* "Acoustic Behavior of Animals" (R.-G. Busnel, ed.) pp. 583–654, Elsevier Publ. Co., New York.

Evans, W. E., and Harmon, S. R., 1968, Experimenting with trained pinnipeds in the open sea, *in* "The Behavior and Physiology of Pinnipeds" (R. J. Harrison, R. C. Hubbard, R. S. Peterson, C. E. Rice, and R. J. Schusterman, eds.) pp. 196–208, Appleton-Century-Crofts, Inc., New York.

Fish, J. F., 1966, Sound production in the American lobster, *Homarus americanus* H. Milne Edwards (Decapoda Reptantia), *Crustaceana* **11:** 105–106.

Fish, J. F., 1968, The effect of sound playback on the toadfish (*Opsanus tau*), *Am. Zoologist* **8(3):** 691 (abst.).

Fish, J. F., 1969, The effect of sound playback on the toadfish (*Opsanus tau*), dissertation abst., University Microfilm No. 70–14, 148, 3 pp.

Fish, J. F., and Vania, J. S., 1971, Killer whale, *Orcinus orca*, sounds repel white whales, *Delphinapterus leucas*, *Fish. Bull.* **69(3):** 531–536.

Fish, M. P., 1964, Biological sources of sustained ambient sea noise, *in* "Marine Bio-acoustics" (W. N. Tavolga, ed.) pp. 175–194, Pergamon Press, New York.

Freytag, G., 1968, Ergebnisse zur marinen Bioakustik, *Protok. Fisch Tech.* **11(52):** 252–352.

Frings, H., 1964, Problems and prospects in research on marine invertebrate sound production and reception, *in* "Marine Bio-acoustics" (W. N. Tavolga, ed.) pp. 155–172, Pergamon Press, New York.

Frings, H., and Frings, M., 1967, Underwater sound fields and behavior of marine invertebrates, *in* "Marine Bio-acoustics" (W. N. Tavolga, ed.) Vol. 2, pp. 261–281, Pergamon Press, New York.

Green, J., 1962, "A Biology of Crustacea," Quadrangle Books, Chicago, 180 pp.

Hashimoto, T., and Maniwa, Y., 1967, Research on the luring of fish shoals by utilizing underwater acoustical equipment, *in* "Marine Bio-acoustics" (W. N. Tavolga, ed.) Vol. 2, pp. 93–101, Pergamon Press, New York.

Hazlett, B. A., and Winn, H. E., 1962, Sound production and associated behavior of Bermuda crustaceans (*Panulirus*, *Gonodactylus*, *Alpheus*, and *Synalpheus*), *Crustaceana* **4:** 25–38.

Headquarters, Pacific Missile Range, 1965, News Release/Public Affairs Office, Point Mugu, California, 3 pp.

Hobson, E., 1963, Feeding behavior in three species of sharks, *Pac. Sci.* **17:** 171–194.

Johnson, C. S., 1967, Sound detection thresholds in marine mammals, *in* "Marine Bio-acoustics" (W. N. Tavolga, ed.) Vol. 2, pp. 247–255, Pergamon Press, New York.

Johnson, M. W., 1943, Underwater sounds of biological origin, UCDWR Rep. U28, PB48635, pp. 1–26.

Josephson, R. K., 1961, The responses of a hydroid to weak water-borne disturbances, *J. Exp. Biol.* **38:** 17–27.

Kronengold, M., Dann, R., Green, W. C., and Loewenstein, J. M., 1964, Description of the system, *in* "Marine Bio-acoustics" (W. N. Tavolga, ed.) pp. 11–25, Pergamon Press, New York.

Laverack, M. S., 1962, Responses of cuticular sense organs of the lobster, *Homarus vulgaris* (Crustacea). I. Hair-peg organs as water current receptors, *J. Comp. Biochem. Physiol.* **5:** 319–325.

Lindberg, R. G., 1965, Growth, population dynamics, and field behavior in the spiny lobster, *Panulirus interruptus* (Randall), *Univ. Calif. Publ. Zool.* **59(6):** 157–248.

Møhl, B. 1968, Hearing in seals, *in* "The Behavior and Physiology of Pinnipeds" (R. J. Harrison, R. C. Hubbard, R. S. Peterson, C. E. Rice, and R. J. Schusterman, eds.) pp. 196–208, Appleton-Century-Crofts, Inc., New York.

Moulton, J. M., 1956, Influencing the calling of sea robins (*Prionotus* spp.) with sound, *Biol. Bull.* **111(3):** 393–398.

Moulton, J. M., 1957, Sound production in the spiny lobster *Panulirus argus* (Latr.), *Biol. Bull.* **113(2):** 286–295.

Moulton, J. M., 1960, Swimming sounds and the schooling of fishes, *Biol. Bull.* **119(2):** 210–223.

Moulton, J. M., 1963, Acoustic orientation of marine fishes and invertebrates, *Ergeb. Biol.* **26:** 27–39.

Moulton, J. M., and Backus, R. H., 1955, Annotated references concerning the effects of man-made sounds on the movements of fishes, Dept. Sea and Shore Fisheries, Augusta, Maine, Fish. Cir. No. 17, 7 pp.

Myrberg, A A., Jr., 1968, Present research in elasmobranch biology being carried out by the Behavior Section of the Institute of Marine Sciences, AIBS Shark Research Panel, pp. 95–96.

Myrberg, A. A., Jr., 1969, Glassell Building, General information on salt water and other systems, Inst. Mar. Sci. Rep. Offset, 8 pp.

Myrberg, A. A., Jr., in press, Ethology of the bicolor damselfish, *Eupomacentrus partitus* (Pisces: Pomacentridae). A comparative analysis of laboratory and field behavior, *Anim. Behav. Monogr.*

Myrberg, A. A., Jr., and Banner, A., 1970, Studies on the behavior of sharks with emphasis on their acoustic environment, AIBS Shark Research Panel, pp. 86–88.

Myrberg, A. A., Jr., Stevenson, R. A., and Steinberg, J., 1966, Biological considerations for the future use of the original television housing of the Bimini video-acoustic project, ONR Tech. Rep., 13 pp.

Myrberg, A. A., Jr., Banner, A., and Richard, J. D., 1969, Shark attraction using a video-acoustic system, *Mar. Biol.* **2(3):** 264–276.

Nelson, D. R., 1965, Hearing and acoustic orientation in the lemon shark, *Negaprion brevirostris* (Poey), and other large sharks, Doctoral dissertation, University of Miami, 146 pp.

Nelson, D. R., 1968, Hearing and acoustic orientation in sharks, AIBS Shark Research Panel, pp. 99–100.

Nelson, D. R., and Gruber, S. H., 1963, Sharks: Attraction by low-frequency sounds, *Science* **142(3594):** 975–977.

Nelson, D. R., and Johnson, R. H., 1970*a*, Acoustical studies on sharks, AIBS Shark Research Panel, pp. 89–90.

Nelson, D. R., and Johnson, R. H., 1970*b*, Acoustic studies on sharks, Rangiroa Atoll, 1969, Long Beach California State College, ONR Tech. Rep. No. 2, 16 pp.

Nelson, D. R., Johnson, R. H., and Waldrop, L. G., 1969, Responses in Bahamian sharks and groupers to low-frequency, pulsed sounds, *Bull. S. Calif. Acad. Sci.* **68(3):** 131–137.

Norris, K. S., 1965, Trained porpoise released in open sea, *Science* **147(3661):** 1048–1050.

Richard, J. D., 1968, Fish attraction with pulsed low-frequency sound, *J. Fish. Res. Bd. Canad.* **25(7):** 1441–1452.

Ridgway, S. H., 1966, Studies on diving depths and duration in *Tursiops truncatus*, Proc. 3rd Ann. Conf. Biol. Sonar and Diving Mammals, Stan. Res. Inst. Menlo Park, Calif., pp. 151–158.

Ridgway, S. H., Scronce, B. L., and Kanwisher, J., 1969, Respiration and deep diving in the bottlenose porpoise, *Science* **166(3913):** 1651–1654.

Salmon, M., 1965, Waving display and sound production in the courtship behavior of *Uca pugilator*, with comparisons to *U. minax* and *U. pugnax*, *Zoologica* **50**: 123–149.
Salmon, M., 1967, Coastal distribution, display and sound production by Florida fiddler crabs (Genus *Uca*), *Anim. Behav.* **15**: 449–459.
Salmon, M., and Atsaides, S., 1968, Visual and acoustical signalling during courtship by fiddler crabs (Genus *Uca*), *Am. Zoologist* **8(3)**: 623–639.
Salmon, M., and Stout, J. F., 1962, Sexual discrimination and sound production in *Uca pugilator* Bosc, *Zoologica* **47(1)**: 15–21.
Shiskova, E. V., 1958, Concerning the reactions of fish to sounds and the spectrum of trawler noise (trans. by J. M. Moulton), *Ryb. Khoziai.* **34**: 33–39.
Steinberg, J. C., and Koczy, F. F., 1964, Objectives and requirements, *in* "Marine Bioacoustics" (W. N. Tavolga, ed.) pp. 1–9, Pergamon Press, New York.
Steinberg, J. C., Cummings, W. C., Brahy, B. D., and MacBain (Spires), J. Y., 1965, Further bio-acoustic studies off the west coast of North Bimini, Bahamas, *Bull. Mar. Sci.* **15(4)**: 942–963.
Stevenson, R. A., Jr., 1967, Underwater television, *Oceanology Internat.* **2(7)**: 30–35.
Tavolga, W. N., 1958, The significance of underwater sounds produced by males of the gobiid fish, *Bathygobius soporator*, *Physiol. Zool.* **31(4)**: 259–271.
Tavolga, W. N., 1965, Review of marine bio-acoustics. State of the art: 1964, Tech. Rep. NAVTRADEVCEN 1212–1, 100 pp.
Watkins, W. A., and Schevill, W. E., 1968, Underwater playbacks of their own sounds to *Leptonychotes* (Weddell seals), *J. Mammal.* **49(2)**: 287–296.
Westenberg., J., 1953, Acoustical aspects of some Indonesian fisheries, *J. Cons.* **18**: 311–325.
Winn, H. E., 1967, Vocal facilitation and the biological significance of toadfish sounds, *in* "Marine Bio-acoustics" (W. N. Tavolga, ed.) Vol. 2, pp. 283–303, Pergamon Press, New York.
Winn, H. E., and Marshall, J. A., 1963, Sound producing organ of the squirrelfish, *Holocentrus rufus*, *Physiol. Zool.* **36(1)**: 34–44.
Winn, H. E., Marshall, J. A., and Hazlett, B., 1964, Behavior, diel activities, and stimuli that elicit sound production and reactions to sounds in the longspine squirrelfish, *Copeia* **1964** **(2)**: 413–425.
Wisby, W., 1964, Seawater supply in the tropics, *in* "Seawater Systems for Experimental Aquariums" (J. R. Clark, and R. L. Clark, eds.) U.S. Fish Wildlife Serv., Res. Rep. No. 63, pp. 113–118.
Wisby, W. J., Richard, J. D., Nelson, D. R., and Gruber, S. H., 1964, Sound perception in elasmobranchs, *in* "Marine Bio-acoustics" (W. N. Tavolga, ed.) pp. 255–268, Pergamon Press, New York.
Wolff, D. L., 1966, Akustische Untersuchungen zur Klapperfischerei und verwandter Methoden, *Ztschr. Fisch.* **14(3-4)**: 277–315.
Wood, F. G., Jr., and Ridgway, S. H., 1967, Utilization of porpoises in the Man-in-the Sea program, *in* "Project Sealab Report. An Experimental 45-Day Undersea Saturation Dive at 205 Feet" (D. C. Pauli and G. P. Clapper, eds.) pp. 407–411, ONR Rep. ACR-124.

Chapter 13

VISUAL ACUITY IN PINNIPEDS

Ronald J. Schusterman

Stanford Research Institute
Menlo Park, California
and California State College at
Hayward, Hayward, California

I. INTRODUCTION

Until recently, studies of the sensory adaptations of marine mammals have primarily been carried out with the small-toothed whales and have emphasized hearing mechanisms and acoustic orientation almost to the complete exclusion of the visual organ. The eye, by nature of its large, spatially ordered, point-to-point representation of the environment, often yields instantaneous and panoramic information at considerable distances and is normally considered the most suitable sense organ for environmental orientation. However, previously there was no behavioral evidence of an experimental nature regarding the visual acuity of either the cetaceans or the pinnipeds. Despite experimental evidence suggesting that the behavior of some odontocetes is guided more by sound than by sight (Norris, 1968), Kellogg and Rice (1966) have shown that properly trained dolphins can rapidly solve problems presented to the visual modality, and most recently Spong and White (1969) have done several behavioral experiments concerned with the visual resolving power of the Pacific white-sided dolphin (*Lagenorhynchus obliquidens*) and the killer whale (*Orcinus orca*) underwater. Furthermore, although conditions at the water's surface may indicate extreme darkness below, Hobson (1966) reports that considerably more light may be available to the eye located underwater and has described the tendency of sea lions and seals to approach their prey from below, thus silhouetting the prey against the ambient sky light above.

Unlike the cateceans, which must find food as well as interact socially in an aquatic medium only, pinnipeds, although feeding exclusively underwater, usually spend a good portion of their time resting and interacting socially in a terrestrial environment. Thus, since they are truly amphibious, pinniped forms such as the California and Steller sea lions and the harbor seal require special sensory adaptations, which are probably quite different from those of the more common toothed whales such as the bottlenosed dolphin and the common porpoise.

Although there is abundant evidence indicating that acoustic signalling and orientation on land by several pinniped forms are critical for the establishment and maintenance of their social organization, the role of visual signalling and orientation on land is less well known (Peterson and Bartholomew, 1967). Naturalistic observations have suggested that the visual acuity of the California sea lion on land is quite poor in daylight and is even worse at night (Peterson and Bartholomew, 1967). On the other hand, the role of acoustic orientation by pinnipeds in finding food at sea is not well understood, and it has been suggested by Hobson (1966) and Schusterman (1967) that some forms probably rely a good deal on visual information for this purpose.

In early experiments dealing with the smallest areas of circles and rectangles that seals and sea lions could consistently detect underwater, it was found that the California sea lion (*Zalophus californianus*), Steller sea lion (*Eumetopias jubatus*), and harbor seal (*Phoca vitulina*) were all capable of detecting area differences as small as 6–9 percent and that *Zalophus* was capable of visually detecting differences of 0.25 cm between bars or rectangles (1.5 cm wide) oriented in the vertical plane (Schusterman, 1968). However, these stimuli may have been either totally or partially irrelevant with regard to measuring visual acuity *per se* since the animals could have been gauging merely the total amount of reflected light from around the area of the stimulus targets, thereby resulting in performance that may have reflected a brightness discrimination. With the use of these stimulus configurations, the animals could, in principle, have discriminated between targets even if their ability to detect fine detail was very poor. Furthermore, in some of the experiments, the animals swam up to the stimulus targets and pushed them; thus there was no way of ascertaining the distance at which the animals made their discriminations. These early efforts at obtaining some information on the underwater visual discrimination capabilities of seals and sea lions were successful in maintaining orienting responses and keeping the sensorimotor requirements relatively constant for different species, in maintaining a clear definition of response, and in keeping the water relatively clear. However, they were largely unsuccessful in keeping the animals at a relatively fixed or minimum distance from the stimuli so that accurate visual angles could be calculated, and in using stimulus configurations in which visual resolving power could be measured uncontaminated by intensity discriminations (Riggs, 1965).

To correct these deficiencies, several changes in methodology were made. In one set of experiments, conditioned vocalizations were used as an objctive index of the fact the that animal could discriminate between patterns. In these experiments, the animal was trained to place its head in a head holder so that all responses to the stimuli were made while the head was in a fixed position. In other experiments, the animals were allowed to swim toward and press the stimulus targets; however, a barrier extended out between the stimuli, and the animals were trained to make their decisions at a minimal distance from the acuity targets. In both sets of experiments, the stimulus targets used were a series of grating patterns in which the widths of the lines varied from coarse to fine; the result was that visual acuity, which has been defined as the spatial resolving capacity of the visual system, could be specified in terms of the angular width of the line of the finest grating that could be resolved.

II. UNDERWATER

A. Methodology

In this section I describe a technique used to measure the underwater visual acuity of a 4-year-old male Steller sea lion (Runner) and a 5-year-old male harbor seal (Goldie). Although the general health of both animals was good throughout the training and threshold phases of the study, the corneas of both eyes of the Steller sea lion were slightly "milky," and on a few test sessions during the first threshold phase of the experiments this animal kept its left eye closed while swimming toward the acuity targets. However, binocular and monocular viewing had no effect on the performance of this sea lion. Inspection of the harbor seal's eyes indicated that they were in excellent condition throughout the study. Previously, both animals had received extensive experience with pattern discriminations (the shapes used were painted black on a white background) in which their task was to push one of two targets with the nose in order to receive a fish reward (Schusterman, 1968).

The animals were trained and tested outdoors between the hours of 0800 and 1200 in an oval tank, and were not fed for approximately 20 hr prior to each test session. The tank was constructed of redwood painted white and measuring 4.6 by 9.1 by 1.8 m. Pictures and details of the testing conditions and apparatus have been published (Schusterman, 1967, 1968). The acuity targets were produced from 12.7-cm^2 photos of Ronchi rulings with the black and white stripes of equal width. The *standard* grating consisted of 300 lines per inch (0.05 mm in width). The lines were invisible to the human eye without the aid of a lens and appeared as a flat gray square. The *variable* gratings consisted of lines varying in width from 25.4 to 0.96 mm. All variable gratings used during the final acuity threshold tests were compared with the standard by three human observers at a distance that prevented resolution of the lines, and in all cases observers reported that the variable gratings were indistinguishable from the standard grating. Photos of the horizontal striations were centrally fixed and laminated within a 22.8-cm² clear plexiglass, 0.4 cm thick. The area framing the acuity grating was painted flat black. These plexiglass squares could be slipped in and out of an aluminum frame (see Figs. 1 and 2). Thus, the animals saw a 12.7-cm² acuity grating (black and white stripes) surrounded by a large black border.

Two targets were always presented simultaneously so that they projected below the opaque screen, with the center of the target being 30 cm below the water level. At the beginning of a trial, a stimulus panel located behind the opaque screen was lowered to the water level. Attached to the side of the stimulus panel facing the experimenter were two rods holding the targets. Deflection of either rod activated a microswitch that produced an audible click signal. A perpendicular divider of mesh wire projected down to the floor of the tank and 68.6 cm outward from the opaque screen, thus lying between the targets and preventing the animals from swimming laterally from one target to the other. The distance between the centers of any two targets was 67 cm.

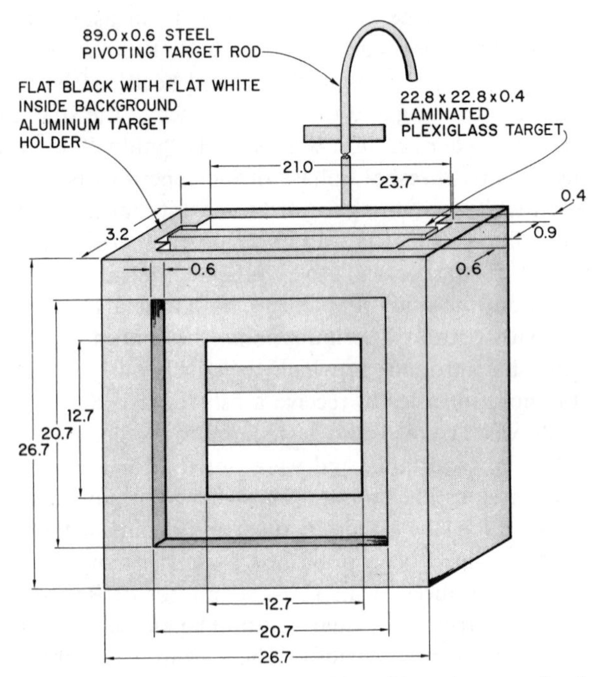

Fig. 1. Schematic diagram of target holder with acuity target in place.

Fig. 2. Underwater photo of harbor seal (left) and Steller sea lion (right) waiting for fish reward after having just pushed acuity target. Note that, at distance photo was taken, gratings cannot be resolved. Head holder with white stand shown in foreground.

The animals had been previously trained to remain at a position about 6 m in front of the stimulus display area until signalled to approach by the sound of the display being lowered into the water. Their task was to push the target with the variable grating (defining a correct response) in order to obtain a fish reward—a 76-g piece of herring (*Clupea pallasi*) for *Eumetopias* and a 16-g piece for *Phoca*. Right and left position of the standard and variable gratings was determined by a Gellerman series, and extreme care was taken to avoid differential auditory, visual, or temporal cues between and during stimulus presentations. Moreover, the variable and standard gratings were changed from one target holder to the other to eliminate any cues associated with the target holders. A trial began when the targets were presented and terminated when they were withdrawn. If the animal, in swimming toward the target, extended its head beyond the outer point of the 68.6-cm-long stimulus divider on the side of the standard grating, then it was forced either to press the target on that side or to swim back to its starting position. If it attempted to switch sides after its head passed the divider, the targets were raised immediately and the response was counted as an "error." In this manner, animals approaching the targets were trained to make their final decision at a minimal distance from the acuity targets, i.e., at the barrier. Thus the minimum distance between the gratings and the animal was taken as 68.6 cm (27 inches), and visual angles were calculated for each variable grating on the basis of this distance.

Two observers were always present throughout the entire study. One presented the stimulus display while the other observed the animal from the testing platform, recording correct responses, response latencies (by means of a stop watch), and the presence of orienting responses, which were defined as postural changes of the head or body occurring within 30 cm of the barrier. Previous estimates of reliability in scoring an orienting response by two independent observers yielded agreement of better than 90 percent (Schusterman, 1965*a,b*).

The experiment consisted of three principal phases. During the first phase, the standard and a variable target with striations 25.4 mm wide were presented to the subjects, who were reinforced only for responding to the variable target. The Steller sea lion showed an immediate preference for the reinforced or positive target; he committed only eight scattered errors during the first 67 trials. The harbor seal, on the other hand, showed a slight preference for the standard target, and it took 110 trials and 48 errors before he consistently responded (30 consecutive correct responses) to the variable target. Stimulus control was then gradually shifted to targets having finer striations. Such gradual shifting of stimulus control from relatively easy to more difficult discriminations within a given stimulus dimension has proved to be an extremely efficient technique in training many different animals, including sea lions, for psychophysical experiments in that it tends to decrease or eliminate emotional behavior and may teach the animal to attend to both the location and nature of the critical cue (Blough, 1966; Schusterman, 1968).

In the second phase, a modified method of limits was used to obtain a range of acuity targets necessary for estimating thresholds (defined as the interpolated values at which the animals responded correctly 75 percent of the time) during the third and final phase of the study. There were a total of 62 sessions for the Steller sea lion and 50 sessions for the harbor seal. Both animals showed significant improvement from the early sessions to the later sessions, with performances stabilizing during the last ten to 20 sessions. At each session a series of acuity gratings was presented, starting with stripes 3.2 mm in width. The stripes were made finer or broader, depending on whether the animal succeeded in making eight or more correct responses in ten successive trials. A session was terminated only after an animal had succeeded at a given acuity grating. The minimum number of trials at a session was 60, and on a few occasions an animal was given as many as 90 trials. By this means, threshold estimates for the two animals were bracketed at between 5.5′ and 8.5′ of visual angle.

During the final phase, an underwater visual acuity threshold for each animal was obtained by the psychophysical method of constant stimuli.

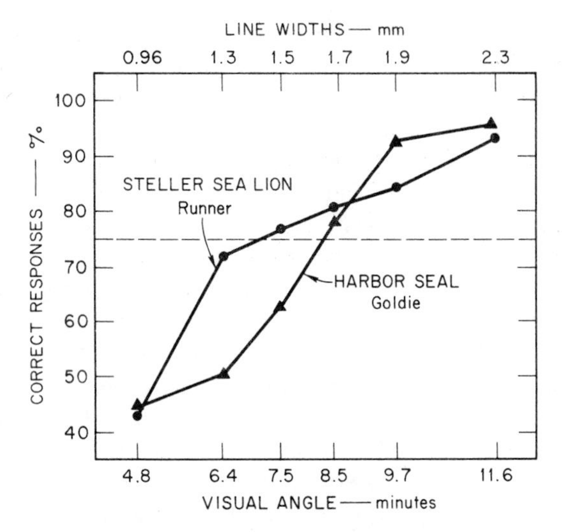

Fig. 3. Correct responses as a function of visual angle calculated at a distance of 68.6 cm.

Following a ten-trial warm-up period with a suprathreshold target, each of six variable gratings with line widths as listed on the top abscissa of Fig. 3 was paired randomly from session to session with the standard for ten consecutive trials, for a total of 70 trials per session. A total of 15 sessions was run for each animal.

Ambient light measurements were taken with an SEI photometer during and immediately after the final threshold phase was completed. Although most of the tank was in sunlight, even on clear days acuity targets were always presented in the shade of the testing platform. Ambient light measurements around the stimulus display area were taken from behind the back window of the tank when it was filled with water, and yielded readings of 130 mL on clear days and 85 mL on overcast days.

B. Results

The underwater visual acuity threshold curves obtained by the method of constant stimuli for one *Eumetopias* and one *Phoca* are shown in Fig. 3. Several comparisons may be made from these curves: (1) *Eumetopias*, despite its slightly "milky" eyes, performed significantly better than chance ($P < 0.01$) with line widths that substended a visual angle of 6.4′ of arc, but *Phoca* did not achieve performance significantly better than chance until tested with broader stripes subtending a visual angle of 7.5′; (2) whereas the acuity curve for *Phoca* showed the typical ogive function resulting from a psychophysical experiment, the curve obtained for *Eumetopias* was negatively accelerated;

(3) both animals were capable of resolving stripes that subtended a visual angle of 8.5′ of arc with relatively good accuracy; (4) threshold estimates measured in minutes of visual angle over the 15 test sessions were 7.1 for *Eumetopias* and 8.3 for *Phoca.*

Threshold values during the first eight and the last seven test days were 7.5′ and 6.5′ for *Eumetopias* and 8.6′ and 8.1′ for *Phoca.* Although there is some evidence of improvement, especially in the case of *Eumetopias*, the differences revealed in these comparisons were not significant and indicate relatively stable acuity performance by both animals during the final threshold phase. This resulted presumably from the rather long and intense training and early threshold phases.

Pinnipeds, while swimming underwater, frequently acquire orienting and fixating responses of the head and eyes in order to discriminate effectively (Schusterman, 1955*a,b*, 1966). Conflict and discrimination difficulty are critical factors in the production and maintenance of orienting responses (Tolman, 1948). On this basis, one would predict that orienting responses by the seal and sea lion during the present experiment would be maintained at a relatively high level when they were attempting to detect acuity targets below or just above threshold as compared to suprathreshold targets. The results, plotting orienting responses as a function of visual angle, are shown in Fig. 4 and bear out this hypothesis. Significantly more orienting responses were made on the most difficult discrimination than on each of the two easiest discriminations ($P < 0.05$, sign test). Predictably, their response latencies were also increased as the stripes of the acuity targets became finer (see Table I), with latencies for both animals being significantly longer on the most difficult discriminations as compared to the easiest discrimination ($P < 0.05$,

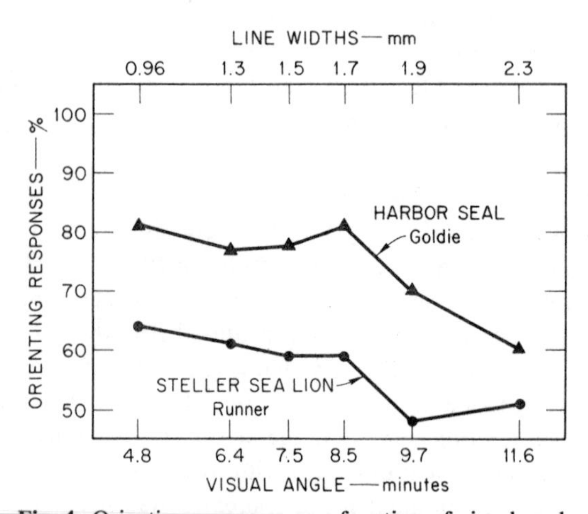

Fig. 4. Orienting responses as a function of visual angle.

Table I. Response Latencies (in sec) as a Function of Pattern Discriminability

Line width (mm)	Visual angle (′)	Harbor seal "Goldie"		Steller sea lion "Runner"	
		Mean	SD	Mean	SD
0.96	4.8	8.1	1.4	6.4	1.9
1.3	6.4	7.7	2.1	6.1	1.5
1.5	7.5	7.8	2.0	5.9	1.0
1.7	8.5	8.1	1.8	5.9	1.6
1.9	9.7	7.0	1.8	4.7	1.5
2.3	11.6	6.0	1.8	5.1	1.1

sign test). The extremely high level of orienting responses as well as the relatively longer latencies on the most difficult discriminations reflect an attempt by both animals to swim as close as possible to the stimulus display area in order to resolve the finer gratings prior to making a decision at the edge of the barrier. Occasionally, in the case of *Eumetopias*, if the animal swam past the edge of the barrier on the side of the standard target, it would swim to within about 50 cm of the target and then turn away from the target and swim back to the start position.

III. AERIAL VS. UNDERWATER

A. Methodology

This section describes several experiments in which a sexually mature male (Sam) and female (Bibi) *Zalophus* were trained to emit a burst of short-duration sound pulses or clicks (see Schusterman, 1968, for a review of training procedures) when viewing targets consisting of black and white stripes and to remain silent when viewing a target that appeared flat gray. By training the animal to keep its head in a fixed position while making the discrimination and by varying the width of the stripes, an estimate was obtained of the finest detail that could be resolved by the sea lion eye. Threshold data were acquired in this manner, both in air and underwater at several different distances, thus allowing for direct comparison of visual acuity in the two media (see Fig. 5).

Testing was outdoors between 0800 and 1200 hr in the oval redwood tank described previously. Animals were not fed for 20 hr before a test. Acuity targets were the same as those used for testing underwater visual acuity in the Steller sea lion and harbor seal.

A "correct" response was defined as either emitting a burst of clicks

Fig. 5. California sea lion viewing variable acuity target in air.

to the variable stimulus or remaining silent in the presence of the standard stimulus. Sam was required to remain silent for a 5-sec period. Bibi was required to remain silent for only a 3-sec period. The change in response for Bibi was necessary because when she remained silent she jerked her head back and forth and moved forward toward the target; these movements only occurred while she was silent. If an animal produced a burst of clicks, it would usually do so within less than 0.5 sec following stimulus presentation, which was always by the successive method (i.e., one at a time). Two experimenters worked from behind an opaque screen, presenting stimuli, recording responses, observing the animal, and reinforcing all correct responses. Sound production was continuously monitored by means of a hydrophone. The same procedure for testing visual acuity was used underwater and in air, the only difference being that in air the water level of the tank was lowered so that the targets were in air and the animal's eyes remained in air while the neck and the rest of its body remained in water. Despite the fact that in the aerial testing the animal's mouth was out of water, its neck was in water, and it is known that vocalizations made by sea lions with their mouths closed and out of water may still be projected under water by the larynx and thereby be picked up by a hydrophone (Schusterman and Balliet, 1969; Schusterman *et al.*, 1970).

Like the previously described study of underwater visual acuity, this experiment consisted of three phases. First, the standard and a variable target with striations 25.4 mm wide were presented, and the sea lions were reinforced for vocalizing and remaining silent in response to the appropriate target. Once stimulus control of vocalization was obtained with this variable target, it was gradually shifted to targets having finer striations.

In the next phase, a modified method of limits was used to obtain a range of acuity targets necessary for estimating thresholds during the final phase of the study. Threshold bracketing by this method was primarily accomplished with the sea lion's head being maintained at a fixed distance of 4.5 m from the targets. However, distances of 1.9 and 5.5 m were also used. It is important to note that both animals showed a significant practice effect *only* during aerial testing. At the beginning of a test session, it was found that performance in air was generally more erratic than it was underwater. Therefore, a worm-up period was used in air, with both animals being required to perform at an 80 percent level of accuracy before proceeding to the more difficult discriminations during a given test session. These early results suggest that *Zalophus* learns to orient to visual patterns slower in air than it does underwater (possibly to adjust or compensate for astigmatic blurring); this may be the reason that naturalists have suggested that the aerial vision of these animals is very poor in daylight (Peterson and Bartholomew, 1967; Hamilton, 1934). When performances stabilized, threshold estimates were bracketed between 5′ and 8′ of visual angle, with little difference between aerial and underwater performance.

During the final phase, acuity thresholds were obtained by the psychophysical method of constant stimuli. Following a ten-trial warm-up period with a suprathreshold target, each of four variable gratings with line widths as listed on the top of Fig. 6 was paired randomly from session to session with the standard for ten consecutive trials and then repeated within each session, for a total of 90 trials per test session. A total of 20 sessions were run, with testing in air and water alternated daily. Thresholds were obtained at three different distances in the following sequence—3.1, 5.5, 1.9, and 5.5 m. Thus two separate thresholds were computed at the disatnce of 5.5 m.

B. Results

Visual acuity threshold curves are shown in Fig. 6. In general, there appears to be relatively little difference between underwater and aerial acuity at the closest and farthest distances measured. However, at the middle distance, 3.1 m, performance is significantly superior for both animals underwater than in air. The poorer aerial acuity at 3.1 m may be accounted for by the fact that the animals were tested by the method of constant stimuli for the first time at that distance. Performances were best at 5.5 m (see Table II).

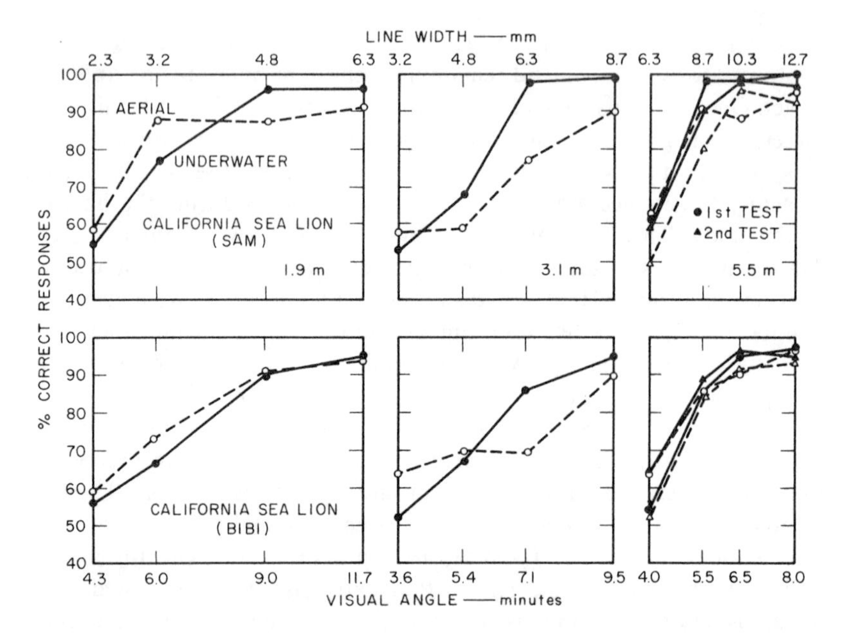

Fig. 6. Comparison of aerial and underwater visual acuity threshold curves at three different distances.

This may be related to latency of vocalization. Although no measurements were made, our general impression was that the animal vocalized more rapidly following stimulus presentations at 1.9 and 3.1 m than at 5.5 m.

Whether or not variable targets were easily distinguished from the standard had little effect on the rate of false alarms (vocal responses to the standard target) by both sea lions. Thus Fig. 7 shows that even though the probability of vocalizing to a variable target (a hit) decreased directly as a function of decreasing visual angles, the probability of making a false alarm remained relatively constant. This means that during difficult discriminations the sea lions did not respond randomly, as one might expect, or persist

Table II. Arieal and Underwater Visual Acuity Thresholds in Minutes of Visual Angle

Distance (m)	Sea lion Sam		Sea lion Bibi	
	Air	water	Air	Water
1.9	5.2	5.8	6.1	7.0
3.1	7.0	5.8	7.8	6.1
5.5 (test 1)	4.8	4.6	4.8	5.0
5.5 (test 2)	5.3	4.8	5.0	4.7
Median	5.25	5.30	5.55	5.55

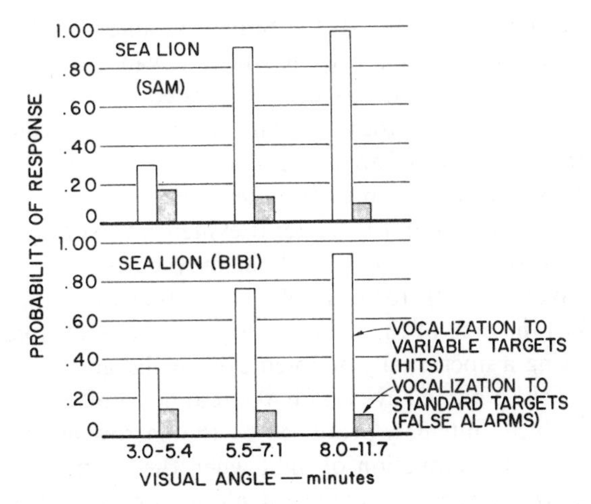

Fig. 7. Relationship between hits and false alarms as a function of visual angle.

in a vocal response. Rather, they did what they were trained to do; i.e., they generally remained silent when viewing a gray patch.

It is well known that without goggles human visual acuity is enormously impaired under water. A preliminary test was made with one human subject, using the same acuity targets previously presented to the sea lions. At a distance of 0.7 m, the subject could quite accurately differentiate the standard target when line widths subtended a visual angle of 7.5′ but could not make such a discrimination when each line subtended a visual angle of 6.2′. At distances of 4.5 and 5.5 m, the subject reported that he could not even see the outline of the targets.

IV. LUMINANCE

A. Introduction

All of the previously described studies on visual acuity have been conducted at relatively high levels of luminance. Since *Zalophus*, as well as several other species of pinnipeds, sometimes procures food in relatively dim light (Hobson, 1966) and appears to be quite active at night on land during the breeding season (Peterson and Bartholomew, 1967), the question arises as to the extent to which underwater and aerial visual acuity of these animals is limited by luminance. This question is of particular importance since it has been suggested that when on land many forms of pinnipeds have become especially dependent upon the acoustic channel for social communication and therefore use their aerial vision very little compared to land mammals

(Peterson and Bartholomew, 1969). Similarly, underwater, where several species including *Zalophus* are quite vocal (Evans, 1967; Poulter, 1968; Schusterman *et al.*, 1970), it has been suggested that when illumination is poor or *almost* nonexistent, these animals, like the odontocete whales, orient to food sources by means of an active sonar system; i.e., they use their own reflected sounds to locate the prey (Poulter, 1966).

There is anatomical and behavioral evidence to suggest that the visual acuity of pinnipeds in extremely low light may be relatively sharp, at least underwater. Walls (1963), for example, states that the number of layers of cells, as well as the area of the tapetum, is greater in seals than in some land mammals having a nocturnally adapted eye. Walls goes on to say that the adaptations of the pinniped eye that make it appear to be nocturnal compensate for the reduced intensity produced by the narrow pupil in air as well as that produced by the reduction of light when the animal dives to relatively great depths. Walls does not comment about how well pinnipeds may be able to see under low levels of luminance except to indicate that with these nocturnal adaptations the animal should have relatively good visual acuity under low levels of luminance while underwater. On the other hand, if the pupil remains constricted in air under low levels of light, then it would be expected that the reduced amount of light should result in poorer visual acuity in air than underwater. However, if the pupil remains dilated in air as well as underwater at very low light levels, then the sea lion should have very poor acuity in air because of the great astigmatism of the cornea (Johnson, 1901; Walls, 1963).

In previous experiments dealing with the effects of luminance on visual acuity, the stimulus configurations used were not appropriate to the visual acuity task *per se*; therefore, it is not surprising that the differences found between the visual acuity of *Zalophus* in air and underwater were not significant, even at levels of luminance down to approximately 10^{-6} mL (Schusterman, 1968, 1969). However, in the present study, which sought to determine aerial and underwater visual acuity thresholds of *Zalophus* under a wide range of luminance, the stimulus configurations used consisted of gratings—the same targets that were used in the previously described experiments.

B. Methodology

The animal tested was a 4-year-old male *Zalophus* (Spike) that had been previously trained and tested on size and line-length discrimination tasks both underwater and in air (Schusterman, 1968). The experiments were conducted in a light-tight tank, which was periodically inspected and corrected for light leaks. Figures 8 and 9 show the principal features of the apparatus, including the target presentation board. At the beginning of each trial, the board

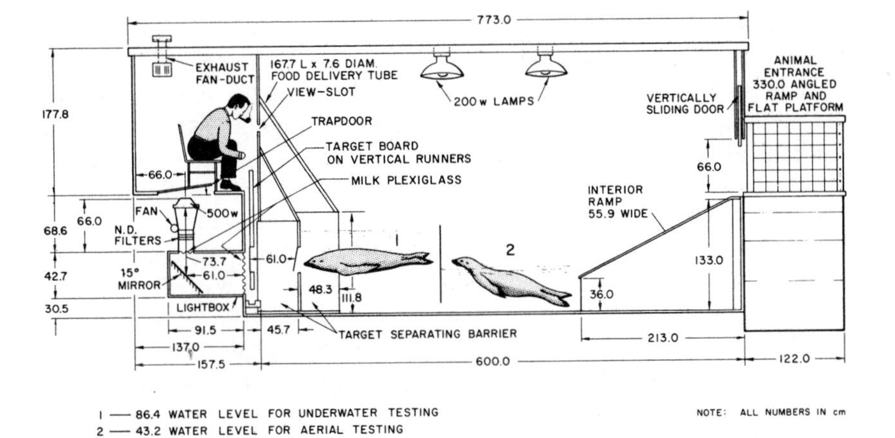

Fig. 8. Schematic diagram of dark tank and apparatus to test sea lion visual acuity under different levels of background luminance. Side view.

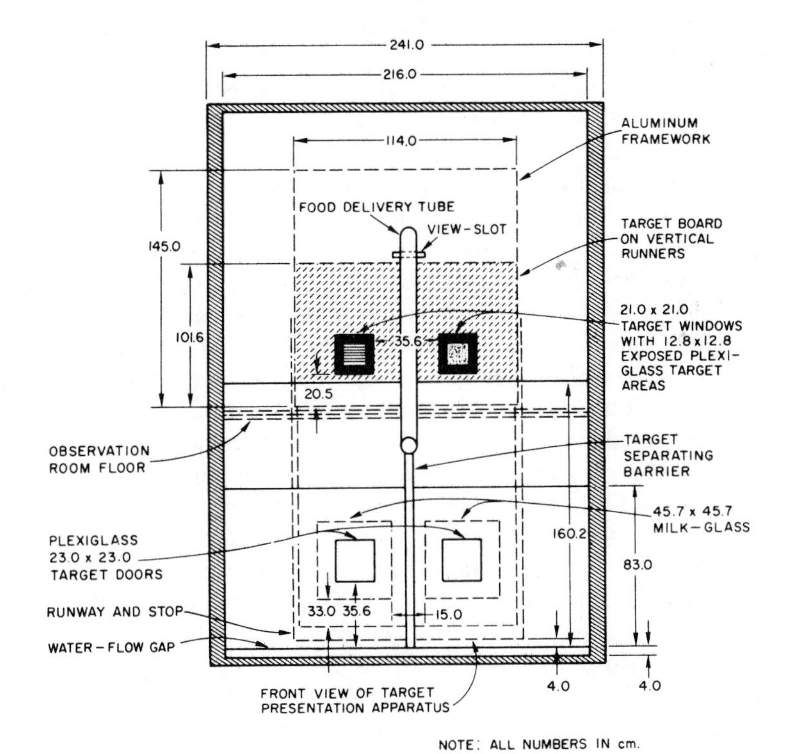

Fig. 9. Schematic diagram of apparatus to test sea lion visual acuity under different levels of background luminance. Front view.

holding the target was lowered from behind an opaque screen. Once the targets were in place, the animal indicated its choice by pressing one of two clear plexiglass windows situated 61 cm in front of the target array. Thus the minimum distance between the acuity gratings and the animal at a constant distance of 61 cm. The response windows were hinged at the top and wired so that a 5-cm displacement activated a microswitch and lit a red 12-W light in the experimenter's compartment. Activation of the microswitch defined the indicator response by the sea lion. An opaque barrier extended out 48.3 cm between the hinged windows and was effective in preventing the sea lion from responding to both windows in rapid succession.

The acuity targets were illuminated from behind by a 500-W General Electric iodine cycle floodlight which was kept at a constant line voltage of 110 and which shone down through a funnel arrangement to a light box. The light was then reflected by a mirror onto two milk plexiglass plates. The intensity of the light falling on the milk plexiglass plates behind the acuity targets was varied with neutral density filters, which were inserted into the funnel arrangement, thus intercepting the light rays before they reached the reflecting surface.

Luminance was calibrated periodically, using either an SEI light meter or a Minolta spot meter. Initially, aerial and underwater measures were taken through a periscope. At the highest intensity used in the experiment, the aerial reading was 0.4 mL higher than the underwater reading.

The animal had been trained to hold at a starting position approximately 4 m in front of the target presentation apparatus. This starting position was near the front of the interior ramp (see Fig. 8). To begin a trial, the experimenter lowered the target board, which released a normally closed microswitch, thus turning on the 500-W lamp. This event signalled the animal to swim forward, either with his head underwater or in air, depending upon the test condition. The animal's task was to strike the window directly in front of the variable target in order to obtain a piece of herring. The experimenter immediately reinforced this response by dropping a piece of herring through the food delivery tube. A response to the standard target was counted as an error and was not reinforced. At the conclusion of the trial, regardless of whether the response was correct or incorrect, the display board was raised. This automatically turned off the 500-W lamp and turned on a 25-W red bulb in the experimenter's compartment, enabling him to record the response and change the position of the target. Position of the targets was randomly determined.

A modified method of limits was used to obtain visual acuity thresholds (defined as the interpolated values at which the animal responded correctly 75 percent of the time) for each of six different luminance levels, ranging from 3 mL to 3×10^{-6} mL. Prior to the start of all test sessions in which limunance was either 3×10^{-2} mL or less, the animal was dark-adapted for 30 min. Dark adaptation was carried out for only 10 min when luminance was at 3 mL. At each test session, the luminance remained constant, and stripes were made finer if the animal succeeded in making eight or more correct responses in ten successive trials, and made broader if the animal committed three or more errors in ten successive trials, Thus, acuity gratings were changed every ten trials. A session was terminated only after the animal had succeeded at a given acuity grating. The minimum number of trials at a given test session was 60, and on a few occasions the animal was given as many as 120 trials. The same neutral density filter was used on two successive test sessions, first in air and then for testing underwater. After obtaining a threshold at a given luminance level, it was decreased by either 1 or 2 log units. Following the first determination of the luminance function, a complete replication was obtained.

C. Results

The main results of this experiment are shown in Table III and Fig. 10. Table III contains the total number of trials and the percentage of correct responses for all visual angles tested under each of the six levels of background lighting. Figure 10 plots aerial and underwater visual acuity thresholds in terms of visual angle as a function of the six levels of background luminance. Both aerial and underwater visual acuity decreased (i.e., targets subtending small visual angles became increasingly difficult to resolve) with decreasing background luminance. However, these luminance functions for visual acuity were dramatically affected by whether the sea lion was viewing the targets underwater or in air. Underwater, visual acuity dropped quite slowly between 3 and 3×10^{-4} mL, and even under the very dim conditions

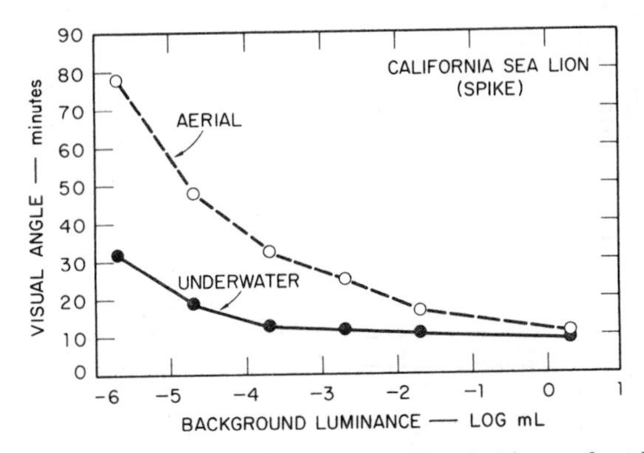

Fig. 10. Aerial and underwater visual acuity thresholds as a function of background luminance.

Table III. Correct Responses of California Sea Lion Spike as a Function of Visual Angle and Luminance[a]

Line width (mm)	Visual angle at 61 cm	Luminance (mL)											
		3		3×10^{-2}		3×10^{-3}		3×10^{-4}		3×10^{-5}		3×10^{-6}	
		No. trials	Percent correct	No. trials	Percent correct	No. trials	Percent correct	No. trials	Percent correct	No. trials	Percent correct	No. trials	Percent correct
						Underwater							
10.3	58.2′											10	100
8.7	47.8′											60	90
6.3	36.3′									60	97	120	83
4.8	26.5′					20	100	20	100	150	92	70	66
3.2	18.0′			40	98	50	96	100	95	150	73	20	45
2.3	13.0′			80	93	110	92	110	75	50	50		
1.9	11.0′	20	100	90	77	80	58	40	58				
1.7	9.6′	100	93	50	66	30	67						
1.5	8.6′	90	60	10	50	10	50						
1.3	7.2′	30	53										
0.96	6.7′	10	50										
						In air							
25.4	2°23.0′											60	97
19.1	1°47.5′									40	95	110	89
12.7	1°11.5′									70	87	100	72
10.3	58.2′									110	84	60	67
8.7	47.8′					10	100	50	94	110	75	20	50
6.3	36.3′			20	100	90	94	140	84	40	60		
4.8	26.5′			80	88	130	80	90	61				
3.2	18.0′	70	97	110	80	60	55	10	50				
2.3	13.0′	90	83	60	45								
1.9	11.0′	100	77										
1.7	9.6′	50	60										
1.5	8.6′	10	40										

[a] All visual angles were calculated at a distance of 61 cm.

a target subtending a visual angle of 13′ could still be resolved by the sea lion in luminance as low as 3×10^{-4} mL. In contrast, the aerial visual acuity of the sea lion, although nearly the same as that underwater at the highest luminance tested (3 mL), dropped precipitously between 3×10^{-3} and 3×10^{-6} mL. Thus, even targets subtending as wide a visual angle as 18′ to 27′ could not be resolved by the sea lion in air in the dim light of 3×10^{-4} mL.

V. SUMMARY AND DISCUSSION

A series of behavioral experiments was conducted using three California sea lions, one Steller sea lion, and one harbor seal to determine the extent to which these pinniped forms can visually resolve fine detail under a variety of conditions. The task for each of the animals was to give a differential response to a striped or a gray target of equal brightness. All targets consisted of lines or grating patterns in which the widths of the black and white lines were equal, regardless of the number of lines. The task of resolving such gratings is generally recognized as one of the most critical aspects of visual acuity, and is less subject to errors involving intensity discriminations than most other types of tasks used to measure visual acuity (Riggs, 1965). In all the experiments conducted, visual angles were readily computed since the animals were always kept at a given or known distance from the acuity targets prior to making a differential response.

The effects of several variables on pinniped visual acuity were studied, both singly and in combination; these included aerial vs. underwater viewing, distance from target, and background luminance. The effects of these variables proved relatively minimal with the exception that luminance had a profound differential effect on aerial and underwater visual acuity in the California sea lion.

In the luminance range of approximately 3–130 mL, all members of the three species studied were capable of consistently resolving gratings subtending visual angles of from 5′ to 9′ of arc underwater in a variety of testing situations. Although these visual acuity thresholds in terms of visual angle are considerably inferior to many of the primates (man, 26″; chimpanzee, 28″; rhesus monkey, 34″), in terms of the currently available comparative data on visual acuity functions, these pinnipeds compare quite favorably with the several species of land mammals reputed to have rather sharp aerial vision, such as the elephant (10′), antelope (11′), red deer (9.5′), and the domestic cat (5.5′) (see Rahmann, 1967, for a review of the literature on comparative visual acuity).

The finding that all three sea lions could resolve lines subtending visual angles of less than 7′, whereas *Phoca* could not, suggests that possibly the visual capability of pinnipeds underwater may differ significantly at the family level and may be related to different feeding or social behavior patterns. However, further study on several additional species with a larger number of subjects is necessary before any such generalizations can be made.

The results, indicating that the visual acuity of the California sea lion in daylight is similar underwater and in air and that aerial acuity is not deleteriously affected at a distance of 5.5 m, *do not support* the view that the eye of pinnipeds is especially adapted for seeing underwater only and that in air the vision of these amphibious marine mammals is subject to gross dioptric errors, including myopia and astigmatism. It is true, however, that all *Zalophus* subjects initially had difficulty resolving grid lines in air and that special "warm-up" sessions had to be instituted before their aerial acuity matched their underwater acuity. This suggests that, in air, pinnipeds make special postural and visual accommodative adjustments in order to fix a relatively clear image on the retina. Aerial and underwater comparisons of the visual acuity in *Zalophus* under good or moderate lighting conditions are supportive of the stenopaic theory of pinniped aerial visual acuity (Walls, 1963; Johnson, 1901). According to this theory, underwater—where the cornea plays no role in focusing an image on the retina—the relationship between the position of the retina and the refractive strength of the lens is such that the animal will have a sharp retinal image. On the other hand, in air—where the cornea makes the eye strongly myopic and astigmatic—the pupil closes down to a narrow vertical slit (stenopaic vision). Therefore, the refractive power of the cornea in the short axis of the slit is irrelevant since the narrow width acts as a pinhole, thus providing the eye with a huge depth of focus in that meridian. Although the optics of the cornea does play a role in the axis parallel to the length of the slit, the astigmatism in that axis combined with the spherical power of the cornea makes the eye approximately emmetropic in that meridian.

Behavioral evidence from one California sea lion on the ability of pinnipeds to resolve detail under low levels of background luminance in air and underwater strongly suggests that decreasing luminance down to 10^{-4} mL has only a slight effect on underwater visual acuity but has a most profound, deleterious effect on aerial visual acuity. At this light level, the aerial vision of *Zalophus* is somewhat poorer than that of the rat under moderate luminance (20–36′—see Rahmann, 1967). but their underwater vision is nearly as good as that of the rhesus monkey and babboon in air (8–13′) at 10^{-4} to 10^{-5} mL (Behar, 1968). With decreasing background luminance, the relative rate of decline in the underwater visual acuity of the sea lion is considerably less than that of humans and several nonhuman primate forms (Behar, 1968). These findings on the sea lion are consistent with anatomical evidence suggesting that the highly specialized "pinniped eye" is structurally adapted for efficient functioning in dim light. However, this is true only with regard to their underwater vision. Their dioptric and accommodative mechanisms appear to be such that although they are adaptive for seeing in air under moderate levels of luminance, they are not adaptive for seeing detail under low levels of luminance, either because the pupil remains constricted in air under low levels of light, thus reducing the amount of light entering the eye, or because the pupil becomes fully dilated in dim light in air, resulting in a significant corneal astigmatism.

It is particular significant to note that the present behavioral results are consistent with the present knowledge about the retinal structure and organization of some pinniped eyes. A recent histological analysis (Landau and Dawson, 1970) of retinas from the California sea lion, the northern fur seal, and the harbor seal revealed no area centralis in any of these pinniped forms, with only rod-shaped receptors being observed. They determined that the ratio of these receptors to ganglion cells was approximately 100:1, a figure similar to estimates of human parafoveal receptor-to-ganglion cell ratios. Landau and Dawson note that visual acuity in rod monochromat humans is similar to the visual acuity of the cat and several pinniped forms, all of which have quite similar recepter-to-ganglion cell ratios.

Because there is not enough information available about the marine ecology of *Zalophus*, *Eumetopias*, and *Phoca*, it is difficult at this time to draw any definite conclusions regarding the ways in which these three pinniped forms use their vision underwater for general orientation, feeding, and social communication. All three species emit vocal signals underwater as well as in air (Schusterman and Balliet, 1969; Schusterman *et al.*, 1970), and there is strong experimental evidence that underwater acoustic signalling plays a role in the social behavior of *Zalophus*. Furthermore, depending on the temporal and geographic factors, the degree of water turbidity may place severe limitations on the visually guided feeding behavior of these animals. This would be especially true of *Phoca*, which sometimes lives in rather muddy bays and estuaries (Scheffer, 1958). Whether *Phoca*, when searching for prey under such conditions, relies on its passive hearing abilities, which have been experimentally shown to be more than reasonably adequate for detecting and localizing sounds underwater (Møhl, 1968), or on echolocation or some other sensory system is still problematic (Poulter, 1968; Schevill *et al.*, 1963). Our results, however, demonstrate that underwater visual resolution in all three species is similar to that of some visually active carnivores on land, such as the cat, and is quite suitable for detecting even relatively small food prey at some distance. The aerial vision of these amphibious mammals also appears well suited for the detection of predators, such as man, as well as for recognizing conspecific individuals or classes of individuals and for recognizing landmarks for purposes of migration. It is likely that many species of sea lions as well as some seals appear to have poor visual acuity in air under daylight conditions, especially with regard to their reactions to human observers, not because their visual acuity capacity is poorly developed for seeing in air in terms of dioptric and accommodative mechanisms and retinal anatomy and organization, but because of selective attention factors. I have suggested elsewhere (Schusterman, 1968) that the reason the aerial visual discrimination of sea lions in daylight appears to be inferior to that of harbor seals is because of the great amount of vigilance behavior shown by harbor seals in the wild and in captivity. The harbor seal, in contrast to sea lions and some other seals, is constantly searching its environment by means of vision and audition, and is therefore much more likely to detect subtle environmental changes.

The present results indicate the potential significance of the visual channel in pinnipeds for such behavioral functions as feeding, predator detection, social communication, and navigation. As has been pointed out elsewhere, each sensory modality has its own special advantages and disadvantages (Marler and Hamilton, 1966). It is likely that pinnipeds as well as other marine mammals use the acoustic or visual channel as receptors, depending upon the situation, or use both channels by combining them in a complementary fashion.

To my knowledge, these are the first visual acuity threshold estimates obtained on any marine mammals using acuity gratings. In earlier experiments, I used a minimum separable discrimination, with patterns consisting of broken and solid black rectangles, 1.5 cm wide and 10 cm long, on a white background. With these test patterns, the minimum discriminable visual angles were estimated to be slightly less than 2′ for both *Phoca* and *Zalophus*. Moreover, there was virtually no difference between the aerial and underwater visual acuity of *Zalophus*, even with background luminance as low as 10^{-6} mL. Using a series of discriminations consisting of two such rectangles versus one, with the space between the two rectangles being progressively reduced, Spong and White (1969) found a minimum discriminable visual angle of approximately 6′ at a distance of 46 cm in two cetaceans, the Pacific white-sided dolphin (*Lagenorhynchus obliquidens*) and the killer whale (*Orcinus orca*). Width and height dimensions of the rectangles used during the final threshold phases of the experiments were 1.3 cm for each of the two rectangles and 2.5 cm for the one rectangle, with the height of the rectangles remaining at 31 cm. In a recent discussion of the types of test patterns and tasks used for measuring visual acuity, Riggs (1965) points out that the resolution of only *two* fine lines depends on the dimensions of the lines, among other things. Broadening of the lines would tend to lower the minimum discriminable threshold, and visual acuity measured in this way probably involves a brightness or intensity discrimination. In view of these considerations and in light of the present results, it seems quite likely that with the types of test patterns used in our earlier attempts to obtain underwater visual acuity thresholds in pinnipeds, the obtained threshold values considerably overestimated their visual resolving power. Perhaps a similar criticism may be leveled at the only experiments thus far reported on the visual acuity of cetaceans (Spong and White, 1969).

ACKNOWLEDGMENTS

This paper is dedicated to the memory of a fellow pinniped observer, Richard S. Peterson. Preparation of this chapter and the research reported herein were supported by Grant GB-7039 from the National Science Foundation and by Contract NOOO14-68-C-0374 from the Office of Naval Research. I thank Richard F. Balliet, Ronald G. Dawson, and Stanley St. John for assistance. A particular note of gratitude is due Dr. T. N. Cornsweet for his helpful suggestions throughtout the course of the research and the manuscript preparation. I thank Roberta for her understanding and encouragement.

REFERENCES

Behar, I., 1968, Visual acuity as a function of luminance in three catarrhine species. Paper presented at 60th Ann. Meet. South. Soc. Philos. and Psychol., Louisville, Ky.

Blough, D. S., 1966, The study of animal sensory processes by operant methods, *in* "Operant Behavior: Areas of Research and Application" (W. K. Honig, ed.) pp. 345–379, Appleton-Century-Crofts, New York.

Evans, W. E., 1967, Vocalization among marine mammals, *in* "Marine Bio-Acoustics" (W. N. Tavolga, ed.), Vol. 2, pp. 159–186, Pergamon Press, New York.

Hamilton, J. W., 1934, The southern sea lion, *Otria byronia* (de Blainville), *Discovery Rep.* **8**: 269–318.

Hobson, E. S., 1966, Visual orientation and feeding in seals and sea lions, *Nature* **210**: 326–327.

Johnson, L., 1901, Contributions to the comparative anatomy of the mammlian eye, chiefly based on ophthalmoscopic examination, *Phil. Trans. Roy. Soc. London Ser. B* **194**: 1–82.

Kellogg, W. N., and Rice, C. E., 1966, Visual discrimination and problem solving in a bottlenose dolphin, *in* "Whales, Dolphins, and Porpoises" (K. S. Norris, end.) pp. 731–754, University of California Press, Berkeley and Los Angles.

Landau, D., and Dawson, W. W., 1970, The histology of retinas from the order Pinnipedia, *Vision Res.* **10**: 691–702.

Marler, F. R., and Hamilton, W. J., III, 1966, "Mechanisms of Animal Behavior," Wiley, New York.

Møhl, B., 1968, Hearing in seals, *in* "The Behavior and Physiology of Pinnipeds" (R. J. Harrison, R. C. Hubbard, R. S. Peterson, C. E. Rice, and R. J. Schusterman, eds.) pp. 172–195, Appleton-Century-Crofts, New York.

Norris, K. S., 1968, The echolocation of marine mammals, *in* "The Biology of Marine Mammals," (H. T. Andersen, ed.) pp. 391–423, Academic Press, New York.

Peterson, R. S., and Bartholomew, G. A., 1967, "The Natural History and Behavior of the California Sea Lion," Am. Soc. Mammologists, Stillwater, Oklahoma.

Peterson, R. S., and Bartholomew, G. A., 1969, Airborne vocal communication in the California sea lion, *Zalophus californianus*, *Anim. Behav.* **17:** 17–24.

Poulter, T. C., 1966, The use of active sonar by the California sea lion, *Zalophus californianus*, *J. Aud. Res.* **6:** 165–173.

Poulter, T. C., 1968, Marine mammals, *in* "Animal Communication," (T. A. Sebeok, ed.) pp. 405–465. Indiana University Press, Bloomington.

Rahmann, H., 1967, Die Sehscharfe bei wirbe Hieren, *Nat. Rdsch.* **20:** 8.

Riggs. L. A., 1965, Visual acuity, *in* "Vision and Visual Perception," (C. H. Graham, ed.) pp. 321–349, Wiley, New York.

Scheffer, V. B., 1958, "Seals, Sea Lions and Walruses. A Review of the Pinnipedia," Stanford University Press, Palo Alto, Calif.

Schevill, W. E., Watkins, W. A., and Ray, C., 1963, Underwater sounds of pinnipeds, *Science* **141:** 50–53.

Schusterman, R. J., 1965*a*, Orienting responses and underwater visual discrimination in the California sea lion, *Proc. 73d Ann. Conv. Am. Psychol. Assoc.* **1:** 139–140.

Schusterman, R. J., 1965*b*, Errorless reversal learning in a California sea lion, *Proc. 73d Ann. Conv. Am. Psychol. Assoc.* **1:** 141–142.

Schusterman, R. J., 1966, Serial discrimination-reversal learning with and without errors by the California sea lion, *J. Exp. Anal. Behav.* **9(5):** 97–104.

Schusterman, R. J., 1967, Perception and determinants of underwater vocalization in the California sea lion, *in* "Les Systemes Sonars Animaux" (R.G. Busnel, ed.) pp. 535–617, Laboratoire d'Acoustique Animale, Jouy-en-Josas, France.

Schusterman, R. J., 1968, Experimental laboratory studies of pinniped behavior, *in* "The Behavior and Physiology of Pinnipeds" (R. J. Harrison, R. C. Hubbard, R. S. Peterson, C. E. Rice, and R. J. Schusterman, eds.) pp. 87–171, Appleton-Century-Crofts, New York.

Schusterman, R. J., 1969, Aerial and underwater visual acuity in the California sea lion as a function of luminance, Naval Undersea Res. and Develop. Center Final Rep, June 5, 1969, Contract N00123–69–C–0208.

Schusterman, R. J., and Balliet, R. F., 1969, Underwater barking by male sea lions (*Zalophus californianus*), *Nature* **222:** 1179–1181.

Schusterman, R. J., Balliet, R. F., and St. John, S., 1970, Vocal displays under water by the gray seal, the harbor seal and the Steller sea lion, *Psychon. Sci.* **18**:303.

Spong, P., and White, D., 1969, Cetacean research at the Vancouver Public Aquarium, Div. Neurol. Sci., Univ. British Columbia Tec. Rep. No. (Summer, 1969), p. 49.

Tolman, E. C., 1958, Cognitive maps in rats and men, *Psychol. Rev.* **55:** 189.

Walls, G. L., 1963, "The Vertebrate Eye," Hafner Publishing Co., New York.

INDEX

This is a joint index for Volumes 1 and 2. Pages 1–244 will be found in Volume 1 and pages 245–492 will be found in Volume 2.

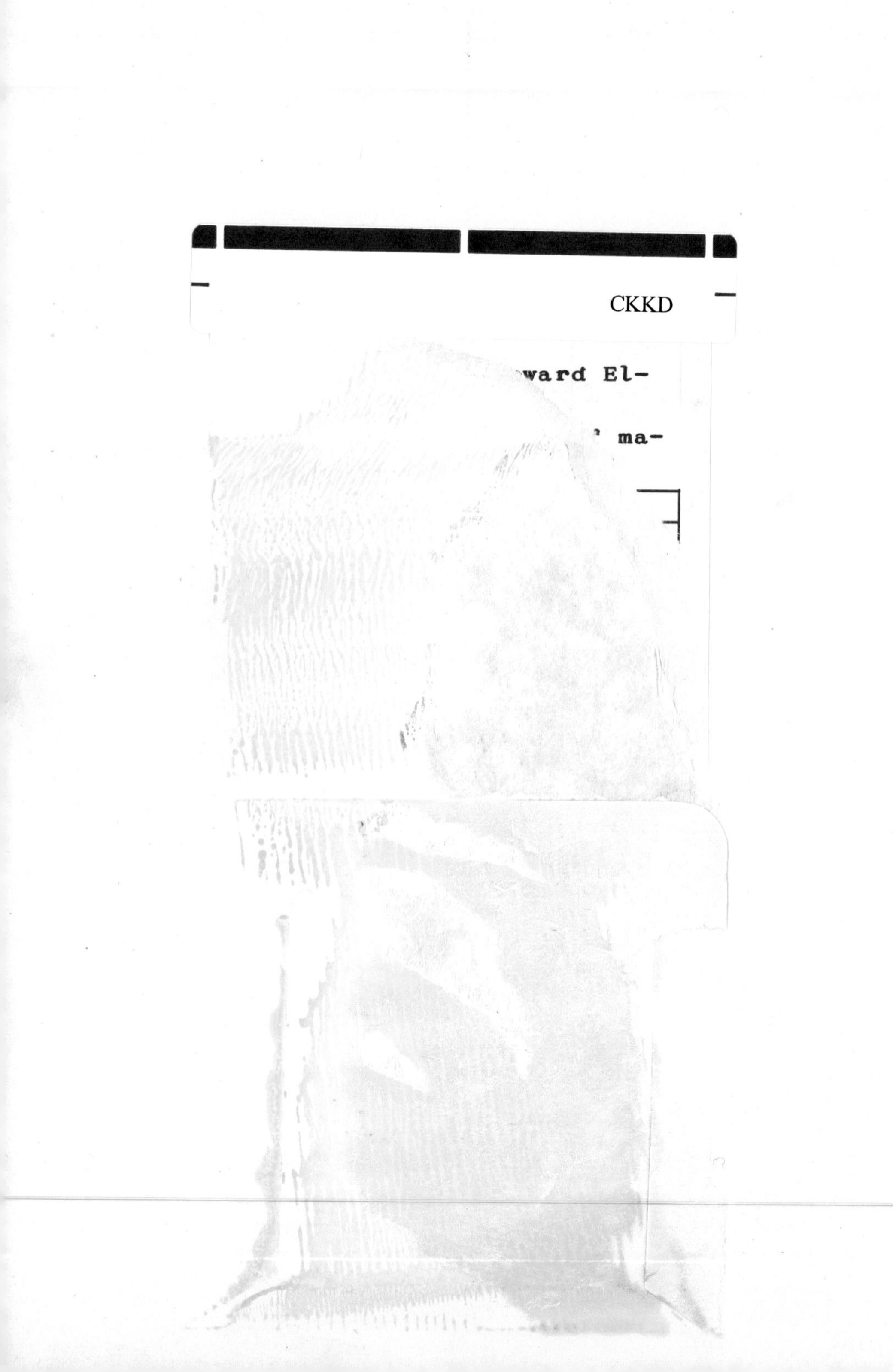
CKKD
ward El-
ma-